Combustion Technology

Vasudevan Raghavan

Combustion Technology

Essentials of Flames and Burners

Second Edition

Ane Books
Pvt. Ltd.

Vasudevan Raghavan
Department of Mechanical Engineering
Indian Institute of Technology Madras
Chennai, Tamil Nadu, India

ISBN 978-3-030-74623-0   ISBN 978-3-030-74621-6   (eBook)
https://doi.org/10.1007/978-3-030-74621-6

Jointly published with ANE Books Pvt. Ltd.
In addition to this printed edition, there is a local printed edition of this work available via Ane Books in
South Asia (India, Pakistan, Sri Lanka, Bangladesh, Nepal and Bhutan) and Africa (all countries in the
African subcontinent).
ISBN of the ANE Books edition: 978-1-119-24178-2

This Springer imprint is published by the registered company Springer Nature Switzerland AG
The registered company address is: Gewerbestrasse 11, 6330 Cham, Switzerland

Preface to Second Edition

I am happy to bring out the second edition of the book on Combustion Technology— Essentials of Flames and Burners. In this edition, at the end of the chapter exercise problems have been added to Chaps. 2, 3, 4, 5 and 6. These problems would provide hands-on training for estimating stoichiometric air, product composition, flame temperature, rate of reactions, flame speed, flame length and other flame characteristics, and calculations involving burners fueled by gas, liquid and solid fuels.

A new chapter (Chap. 8) on numerical modeling of laminar flames has been added. This chapter systematically presents the governing equations for a reacting flow. Terms involved in the governing equations are explained in detail. Calculation of diffusion velocity in a comprehensive manner and simplifications made in the calculations of ordinary and thermal diffusion coefficients have been discussed. All the terms in the energy equation for a reacting flow have been explained thoroughly. Procedures for calculating thermo-physical properties such as density, viscosity, specific heat, thermal conductivity, mass and thermal diffusion coefficients, as a function of temperature and species concentration, have been presented systematically. Various boundary conditions involved in a reacting flow, such as inlet, wall, outlet and free boundary, have been explained in detail with respect to handling primitive variables like velocity, temperature and species mass fractions at these boundaries. The interface boundary conditions required for simulating heterogeneous flames formed over liquid and solid fuels have also been discussed. A section on radiation and soot models has been presented. Here, the basic steps involved in soot formation, and simplified radiation and soot models, have been discussed. A section on mesh and time-step selection has been presented subsequently. Methodology for grid independence study has been explained. The metrics used for checking the convergence have also been discussed. Finally, a case study involving a non-premixed flame of Liquefied Petroleum Gas (LPG) and air has been presented. Realistic composition of LPG with around 12 species has been considered. Results from the simulation using a short chemical kinetic mechanism, and simplified radiation and soot models, have been presented.

I believe that this book can be a good quick reference for graduate and postgraduate students in their first-level combustion courses involving burners, for combustion researchers for comprehending the research areas and for practicing engineers for recalling the necessary basics. This edition brings in the basic concepts of numerical modeling of a reactive flow.

For this edition, I wish to thank my former Ph.D. scholar, Dr. S. Muthu Kumaran, who has simulated the case study problem for Chap. 8 and also proofread that chapter.

Chennai, India   Vasudevan Raghavan

Preface to First Edition

I am happy to bring out this book on Combustion Technology - Essentials of Flames and Burners. This book is mostly based on my lecture notes for post-graduate courses such as Fundamentals of Combustion and Combustion Technology that I teach at IIT Madras. This book is intended for senior under-graduate and beginning post-graduate students in Mechanical and allied engineering disciplines, combustion researchers, and practicing engineers working in the field of combustion. The book concentrates on the essential physical descriptions of the concepts using little mathematical support where required. This book is intended to provide a summary and review of the important concepts related to combustion science and a detailed discussion of the design aspects of different types of flames and burners.

The basic definitions relating to combustion phenomena, descriptions of various types of fuels and their important properties, different modes of combustion, associated emissions, and their consequences on the environment have been presented in the introductory Chapter 1. In Chapter 2, a comprehensive review of the fundamentals aspects of combustion has been presented. This includes the applications of first and second laws of thermodynamics to combustion and the basic aspects of chemical kinetics. Physical descriptions of premixed and non-premixed flames and the aspects of burning of liquid and solid fuel particles are extensively reviewed in Chapter 3. The contents of the first three chapters, therefore, form a comprehensive review of the fundamental aspects of the combustion science.

The next three chapters deal with the applications of combustion phenomena. A detailed discussion of the basic ideas and design features of burners for gaseous fuels is presented in Chapter 4. Important features of three basic gas fuel burners, namely, a co-flow burner, a swirl burner and a naturally entrained air burner, have been presented in detail. Design parameters, performance and emission characteristics have also been discussed with examples. Characteristics of different types of burners for liquid fuels have been presented in Chapter 5. Basic features of different types of atomizers, spray characteristics and spray combustion have also been presented in this chapter. Burners for solid fuels such as coal and biomass are discussed in detail in Chapter 6. Important features of different types of solid fuel burners are discussed with examples. Aspects of gasification of solid fuels using different types of gasification systems are also given in Chapter 6. Finally, a chapter on alternative fuels has been included to bring out the need, characterization and performance of alternative fuels.

Review questions have been provided at the end of each chapter. These questions help the reader to evaluate their understanding of the important concepts covered in that chapter. Several standard text books have been cited in the chapters and are listed towards the end as suggested reading, to enable the readers to refer them when required.

I believe that this book can be a good quick reference for graduate and post-graduate students in their first level combustion courses, for combustion researchers for comprehending the research areas and for practicing engineers for recalling the necessary basics.

Chennai, India   Vasudevan Raghavan

Acknowledgements

I express my heartfelt gratitude to Prof. V. Babu, Department of Mechanical Engineering, IIT Madras, who has motivated me a lot to write this book. He was one of my Ph.D. thesis advisors and has been my best well-wisher and mentor since then. I undertook the task of writing this book only because of his constant urging. I was able to complete writing this book because of his support and encouragement. He has spent an enormous amount of time thoroughly reading the chapters and has made numerous suggestions—both technical and otherwise. As a result of this, the readability of this book has improved significantly. I do not have any words to thank him for his selfless efforts.

I express my sincere gratitude to Prof. T. Sundararajan, Department of Mechanical Engineering, IIT Madras, who was my other Ph.D. thesis advisor and teacher. He has been my well-wisher always. I cannot forget the extra efforts he had put in to conduct many evening classes to explain many concepts related to combustion, in spite of his immense work schedule. He patiently taught me how to go about writing a technical work by bringing out all the associated physics.

My sincere gratitude to Prof. George Gogos, University of Nebraska–Lincoln, USA, who was my mentor during my postdoctoral research career in his laboratory. I have been largely benefitted by the training he gave me on how to conduct rigorous research, thoroughly quantify the research findings and write concisely and clearly.

I sincerely thank my co-worker and friend, Prof. Ali S. Rangwala, Department of Fire Protection Engineering, WPI, USA, for going through the chapters and for providing several useful comments to improve its readability.

My sincere thanks to Prof. U. S. P. Shet, who taught me combustion for the first time when I was a Ph.D. scholar at IIT Madras. He also gave me many ideas on how to construct a laboratory-scale experiment to study laminar flames.

I thank Mr. Bhadraiah, my former M.Tech. student; Mr. Kohli, my former B.Tech. student; and Mr. Sreenivasan, my former M.S. scholar, whose results have appeared in this book. My thanks are due to Dr. Shijin, my former Ph.D. scholar, for simulating a few cases of diffusion flames. He helped me with the numerical results that appeared in the book. I also thank my Ph.D. scholars, Mr. Akhilesh and Mr. Harish, for proofreading the chapters and indicating several corrections.

I sincerely acknowledge the financial support from Curriculum Development Cell of CCE, IIT Madras.

Vasudevan Raghavan

Contents

About the Author

Vasudevan Raghavan is currently working as Professor in the Department of Mechanical Engineering, Indian Institute of Technology Madras (IITM), India. He obtained his Ph.D. degree from IITM and has carried out his post-doctoral research in the University of Nebraska-Lincoln, USA. His areas of research include studies on evaporation and combustion of liquid fuel droplets, computational fluid dynamics applied to reacting flows, laminar flames, fire modelling, flame spread and liquid fuel pool combustion, gasification and combustion of coal and biomass. He has graduated 11 Ph.D. and 18 MS scholars and has authored about 118 international peer reviewed journal articles, 60 international conference articles till date. He teaches graduate courses such as Fundamentals of Combustion, Theory of Fire Propagation, Combustion Technology and Applied Thermodynamics at the Department of Mechanical Engineering in IITM.

Chapter 1
Introduction

In this introductory chapter, basic definitions relating to combustion phenomena are provided. A brief discussion of various types of fuels used in practical applications, along with their important characteristics, is presented next. Following this, the possible modes in which combustion can take place, the emissions generated during combustion of fuels and their consequences on environment are discussed.

1.1 Combustion Process—Basic Definitions

Combustion is an *exothermic oxidation chemical reaction*. In this, a certain quantity of heat is released. The heat release may or may not be accompanied by light emission. In a chemical reaction, there is an exchange of atoms between two reacting molecules as a result of their collision. The number of atoms before and after the reaction remains the same. A combustion reaction involves a *fuel* species and an *oxidizer* species having distinct characters. These species contain atoms such as C, H, O, N and so on. Commonly used fuel species contain C and H atoms, and these are termed as hydrocarbons. Commonly used oxidizer species contain O and N atoms. When a fuel species and an oxidizer species are available in sufficient quantities, along with an ignition source, which is a high temperature region, oxidation of the fuel by the oxidizer takes place. The fuel and the oxidizer are together called *reactants*. As a result of the oxidation process, a set of another species called *product* species are formed and a certain amount of heat is also released. The heat release occurs due to the differences in the energy levels of reactants (having higher energy levels) and the products (having lower energy levels). Depending upon the mode of combustion and type of fuel used, light emission accompanies the heat release.

Examples of fuel species are methane, ethane, ethylene, propane, butane, heptane, benzene, methanol , ethanol, diesel, gasoline, wood, biomass, coal, to name a few.

© The Author(s), under exclusive license to Springer Nature Switzerland AG 2022
V. Raghavan, *Combustion Technology*,
https://doi.org/10.1007/978-3-030-74621-6_1

Some of these fuels are available in gaseous state under normal temperature and pressure, some are available in liquid form naturally and some in solid state. However, combustion reactions predominantly take place in the gaseous state, even though surface reactions in certain solid fuels are known to take place under certain conditions. Therefore, condensed-phase fuels such as liquid and solid fuels have to be gasified before they can participate in the gas-phase exothermic reactions with an oxidizer. This gasification is called vaporization in the case of liquid fuels and pyrolysis in the case of solid fuels. As gas-phase reactions are very fast, transport processes in the case of gaseous fuels and gasification processes in the case of condensed fuels control the overall combustion process.

Selection of a fuel for an application depends upon its availability and the cost, apart from its properties. One of the important characteristics of a fuel is its *calorific value*, which is also called the *heating value*. It is the amount of energy that is released when one kg of the fuel is completely burnt and the products are cooled to standard reference temperature (298 K). The amount of fuel to be supplied is determined based on the calorific value of the fuel and the required heat release, usually expressed on a rate basis as the power rating. For this fuel flow rate, sufficient amount of oxidizer should be supplied in order to burn it completely. The most commonly used oxidizer is the atmospheric air, which is available free of cost, unless there is a specific need to use special oxidizers, which are in general more expensive. Before supplying the atmospheric air to the burner, the moisture content and other dusty impurities present in it are removed. For a given mass flow rate of fuel, a required flow rate of the oxidizer has to be maintained. For example, for 1 kg/s flow rate of methane, at least 17 kg/s of air has to be supplied to ensure almost complete combustion, where it is expected that all the carbon atom in the fuel species is converted to carbon dioxide and all the hydrogen atoms are converted to water vapor. In most of the burner applications, the nitrogen present in the air behaves almost as a chemically inert species and participates only in the transport of heat. Based on the pressure and temperature of the combustion chamber, the type of transport process that takes place and the environmental conditions, there can be several other products, termed as minor products. The usually detected minor products are carbon monoxide, nitric oxides and unburned hydrocarbons. In a few cases, smoke and soot emission are also possible. These minor products are also called pollutants, and it is not desirable to have these in high amounts. There are regulations to restrict the permissible amount of these pollutants that can be produced, depending upon the nature of the application. The accumulation of these pollutants in the atmosphere may be reduced by

(a)   Controlling the transport processes through a proper design of the burner and the combustion chamber.
(b)   Controlling the rates of the chemical reactions that produce them using appropriate catalysts.
(c)   Mitigating the harmful effects of the pollutants that are formed using post-combustion devices such as electrostatic precipitators to remove the pollutants before they are released into the atmosphere.

1.2 Types of Fuels and Their Characteristics

As mentioned earlier, fuels exist in gaseous, liquid or solid states under normal atmospheric pressure and temperature. Fuels may further be classified as *fossil* fuels, formed over several years through geological processes and *synthetic* fuels, which are man-made. Commonly available solid fossil fuels are wood, coal and biomass. Naturally available liquid fuel is crude petroleum oil, and gaseous fuel is natural gas. These fuels are not available in all regions around the globe. Also, depending on the region of availability, the basic properties of the fossil fuels may be different. For instance, coal available in South Africa, Turkey and India has high ash content when compared to the coal available in the USA or Australia. As mentioned in the previous section, for any fuel, its calorific value, or the heating value, is the primary property that governs the feed rate into a combustion system. In addition, properties such as density (or specific gravity), diffusivity and composition (in the case of multi-component fuels) are important for gaseous fuels or for gasified fuels. Few other properties become important for condensed fuels, and these are discussed later.

1.2.1 Gaseous Fuels

Gaseous fuels are more attractive for practical use than condensed fuels since they are cleaner and devoid of ash and mineral particulates. Hence, they can be directly supplied to the combustion chamber. The design of gas burners and the associated transport phenomenon involved are much simpler, as no gasification, atomization and multi-zone chambers are necessary. However, the main disadvantage of gaseous fuels is their low density and the requirement of large storage space (tanks) for storing sufficient amounts of the fuel for practical applications. This disadvantage may be overcome by storing the gas under high pressure in compressed form or by liquefying and storing it under certain conditions. Even by doing this, depending on the application, the quantity of the gas stored may occupy much higher volume when compared to that of a condensed fuel having the same energy value. Further, high-pressure storage demands thick walled pressure vessels and adequate leak proofing measures, which, if not done properly, can cause accidents. A brief look at some commonly used gaseous fuels is presented.

Natural gas is a colorless and odorless gas consisting primarily (around 95% by volume) of methane (CH_4). Traces of higher-order hydrocarbons, carbon dioxide, water vapor and nitrogen are also usually present in natural gas. Some amount of sulfur may also be present. It is available underground over crude oil deposits. It is extracted, compressed and stored in huge storage tanks. In USA and Europe, natural gas is supplied through pipelines for domestic and industrial usage. It is lighter than air, having a molecular mass of around 16 kg/kmol, and it has a calorific value between 50,000 and 55,000 kJ/kg.

Liquefied Petroleum Gas (LPG) is a mixture of various hydrocarbons, with butane and propane as its primary constituents. This evolves during refining of petroleum. When the gas mixture is pressurized to over six times the atmospheric pressure, it liquefies. The liquid is stored in steel cylinders and distributed to homes and industries. Once released through a pressure regulator into the atmosphere, the liquid instantly vaporizes into its gaseous components. LPG has a molecular mass of over 50 kg/kmol and is heavier than air. This has the potential to be an explosion hazard as the gas settles down in the event of a leakage. Commercial LPG has a synthetic odorant added for easy leak detection. It has a calorific value between 45,000 and 50,000 kJ/kg.

Refinery gases are the lighter gases obtained during refining of crude oil. Most of the lighter fractions of the crude oil are converted into gasoline. Other volatile products and lighter gases are extracted as refinery gas. It is a mixture of lighter hydrocarbons, carbon monoxide and hydrogen. These gases are generally used within the refinery itself.

Coal gas is a derived gaseous fuel. This is obtained during carbonization of certain types of coals. The volatiles, which are gaseous components trapped within the coal, are released during this process. These volatiles primarily constitute the coal gas. Steam is also added in certain carbonization process, which helps in converting some of the carbon in the coal to carbon monoxide and hydrogen ($C + H_2O \rightarrow CO + H_2$). Steam also helps in converting CO to H_2 via the water–gas shift reaction, $CO + H_2O \rightarrow CO_2 + H_2$. The composition and calorific value of coal gas will vary based on the rank/type of coal used, and this is discussed in a later section.

Producer gas is obtained from solid fuels in gas generators. It contains by volume 16–20% of carbon monoxide, 16–18% of hydrogen and 8–10% of carbon dioxide, nitrogen and traces of hydrocarbons. Oxygen in the oxidation zone converts C to CO_2 ($C + O_2 \rightarrow CO_2$). This exothermic reaction adds the required heat to the system. When the CO_2 travels through the hot coal layers, CO is formed through $C + CO_2 \rightarrow 2CO$.

Biogas is obtained from biomass such as vegetable and animal wastes. It contains 60–80% of methane and 20–40% of carbon dioxide by volume. Its calorific value lies between 30,000 and 32,000 kJ/kg. Due to the CO_2 content, the calorific values of producer gas and biogas are lower than natural gas and LPG.

Synthetic gas (or syngas) is produced when solid fuels such as coal and biomass are partially burned in insufficient air (less than that required for complete combustion). It primarily consists of CO and H_2. Steam is also used to increase the content of hydrogen in the fuel. The calorific value of syngas depends on the calorific value of the solid fuel, type of gasification process and nature of the oxidizer used.

1.2.2 Liquid Fuels

Liquid fuels are advantageous over gaseous fuels as they have much higher energy density. They are better than the commonly used solid fuels such as coal and biomass, as liquid fuels are relatively cleaner and do not leave ash or minerals as products of combustion. However, unlike gaseous fuels, these fuels have to be gasified or vaporized, before they can participate in the combustion reaction. This calls for few additional properties to characterize a liquid fuel, other than its calorific value. The vaporization is dictated by the *volatility* of the liquid fuel. The volatility is governed by properties such as boiling point, latent heat of vaporization and specific heat. Further, vaporization is a surface phenomenon and surface-to-volume ratio of the liquid dictates the vaporization rate. To increase the surface-to-volume ratio, a liquid fuel jet, injected into a combustion chamber, has to be disintegrated into small droplets by a process called *atomization*. The viscosity of the liquid fuel strongly influences the atomization process. Liquids with lower viscosity can be atomized easily. Furthermore, the ignition of the liquid vapor–air mixture is dictated by the flash and fire points of the liquid fuel. It thus becomes clear that several properties are required to characterize a liquid fuel, which have to be estimated and used in design calculations of liquid fuel burners. These are discussed subsequently.

Flash point of a liquid fuel is the minimum temperature of the liquid at which sufficient vapors are produced. The vapor mixes with atmospheric air and produces a flash or an instantaneous flame, when a pilot flame is introduced over the liquid surface. The flash point corresponds to the formation of an instantaneous premixed flame (formed in properly and already mixed fuel and air) over the fuel surface at a given liquid temperature. When the pilot flame is removed, the flame disappears.

Fire point is a temperature higher than the flash point, at which sufficient vapors are generated. When a pilot ignition source is introduced, a flame is established over the liquid surface and this flame sustains even after the removal of the pilot ignition source. The fire point flame is basically a non-premixed flame, formed around the interface of fuel and air getting mixed in proper proportions.

Boiling point is a temperature higher than the fire point, and it is the saturation temperature of the liquid at the given pressure. Liquids with lower boiling points such as gasoline, methanol, n-heptane and so on vaporize at much rapid rates. Normal boiling point is the saturation temperature at atmospheric pressure. When the liquid reaches its boiling point, there will be no further increase in the temperature as a result of heat addition and all the heat that is added is used to provide the latent heat of vaporization.

The *latent heat of vaporization* is the energy that is required for converting the liquid to its vapor. A higher value of latent heat would indicate that higher amount of heat is required for the phase change at a given pressure and that the liquid may be less volatile.

The liquid fuel is usually at a temperature less than the boiling point to begin with. Heat is required in order to increase its temperature to the boiling point. The heat supplied for this process is termed as *sensible heat*. This depends on the specific heat of the liquid. Then for the phase change, the latent heat has to be supplied. Therefore, the volatility of the liquid fuel is generally governed by the liquid-phase specific heat, boiling point and the latent heat of vaporization.

Further, as mentioned above, the viscosity of the liquid plays an important role in the atomization process. In fact, a liquid can qualify as a fuel, only if its viscosity value

Table 1.1 Boiling and
freezing ranges of gasoline,
kerosene and diesel

Fuel	Liquid density @ 20 °C (kg/m³)	Boiling point (°C)	Freezing point (°C)
Gasoline	710–730	30–80	−80 to −60
Kerosene	775–810	160–200	−60 to −40
Diesel	880–900	220–280	−50 to −20

is within a certain range specified in the standards. Furthermore, one more property that becomes important for a liquid fuel, which is to be used in cold weather, is its *freeze point*.

Liquid fossil fuel, which is usually called *crude oil* or *petroleum*, consists of a large number of hydrocarbon compounds, namely paraffins, isoparaffins, olefins, naphthene and aromatics. Methane, ethane and propane, which are straight chain hydrocarbons, are called normal or saturated paraffins. When the number of carbon atoms is more than four, a compound can exist as a normal paraffin, or with a rearranged structure, as an isoparaffin. Isoparaffin is an isomer of a normal paraffin. Olefins are unsaturated paraffins. Methane, ethane and pentene are a few examples of olefins. Naphthenes are saturated molecules, with a cyclic or ring structure. Compounds which belong to this category usually have their names prefixed with "*cyclo.*" For example, cyclopentane is a naphthene. Even though its chemical formula is C_5H_{10}, due to its cyclic structure, it is saturated. Finally, aromatics are unstructured molecules with a ring structure. Even though they are unstructured, they are more stable than olefins. Crude oil as a whole has by weight 80–85% C, 10–15% H and traces of S, O and N for the remaining proportion.

When crude oil is subjected to fractional distillation process, several components are obtained as products. Gasoline, kerosene and diesel are the primary components used in automotive and aerospace industries. Components heavier than diesel cannot be used in these applications unless processed further. These fuels are multicomponent in nature, and as a result, their density, and boiling and freezing points vary across a range. Table 1.1 shows the boiling and freezing ranges of gasoline, kerosene and diesel, as a function of the liquid density.

Alternative (non-fossil) liquid fuels are alcohols, liquefied solid fuels and vegetable oils. *Ethanol* (ethyl alcohol) is obtained from sources such as sugarcane and corn. Ethanol has characteristics quite similar to gasoline, and it may be blended with gasoline and used directly in engines. At present, a 20–30% blend of ethanol with gasoline is used worldwide in automobiles. Oils extracted from vegetable seeds such as rapeseed, sun flower seed, neem, jatropha and karanja have very high viscosity and cannot be directly used in any spray combustion system, where fuels such as gasoline and diesel can be directly used. Therefore, these straight vegetable oils are subjected to a process called transesterification, using alcohols such as methanol or ethanol. By this process, most of their glycerin content is removed and the oils attain the viscosity values that make them suitable for use as a fuel. This processed vegetable oil with much lesser viscosity is called *biodiesel*. Even though biodiesel has a lower calorific value when compared to diesel, which is its fossil counterpart, its burning characteristics are much cleaner than the diesel; it produces lesser smoke and unburnt hydrocarbon emissions due to the presence of oxygen atoms. At present in unmodified diesel engines, 20% biodiesel–80% diesel blend can be used to reduce the emissions while retaining the same performance. Bio-derived fuels such as alcohols and biodiesels are renewable in nature and hence may be considered to contribute to green energy. However, issues associated with the sustainability of a biofuel, in terms of growing the required amount of the vegetation for its production, have not been resolved completely.

1.2.3 Solid Fuels

Solid fuels, like liquid fuels, have high energy density. Coal, which is a solid fossil fuel, is available in larger quantities when compared to crude oil, a liquid fossil fuel. Wood, which is a common biomass, has been used extensively as a cheap domestic fuel. However, use of wood as a fuel causes deforestation and burning of wood in primitive stoves under certain conditions can be a health hazard owing to the release of carbon monoxide and smoke arising from incomplete combustion. Coal is of plant origin—formed when vegetation buried underground decays and is subjected to high pressures and temperatures over a period of millions of years. Coal is a high-density solid material. It has moisture and gaseous inclusions, which become trapped during its formation. The quality of the coal differs from one location to another based on the degree of completion of the aforementioned formation process.

In general, a solid fuel has *moisture*, gaseous substances called *volatiles*, *fixed carbon* called coke or char and mineral content called *ash*, in some proportions. Proximate analysis is used to determine these fractions. Some solid fuels have sulfur in addition. The calorific value of a solid fuel depends on its volatile and fixed carbon content. The ash content indicates the extent of cleanliness of the solid fuel. A solid fuel is ranked based on its carbon content. Table 1.2 provides the typical mass-based percentages of carbon present in solid fuels. These values are calculated based on a moisture-free, ash-free and sulfur-free basis.

Table 1.2 Typical mass-based percentages of carbon present in various solid fuels based on moisture-, ash- and sulfur-free basis

Fuel	Fixed carbon (%)
Wood	45–55
Peat	50–65
Lignite	60–72
Subbituminous	70–80
Bituminous	80–90
Semi-anthracite	90–95
Anthracite	92–98

The carbon content increases as the formation process near completion and anthracite, with the highest carbon content of more than 92%, is formed toward the end. The volatile percentage decreases as the fixed carbon content increases in the coal. Volatile content in the coal helps in the ignition process. Carbon burning time is much higher and forms the rate-limiting step in coal combustion. Ultimate analysis is used to precisely determine the chemical composition of the coal. Elements such as C, H, N, S, O and *ash* are determined by this analysis.

Rice husk, wheat husk, saw dust and *wood chips* are a few examples of *renewable biomass* fuels. The availability of some of these fuels is seasonal depending on the amount of the harvest. These fuels can be processed and used in heating, cooking and local power generation applications. Since the density of these fuels is much lower than coal and wood, these fuels are often pelletized and used as pellets of various shapes. As in the case with liquid biofuels, the sustainability of these solid biofuels is still a matter of uncertainty. In addition to plant-derived solid fuels, solid fuels can be obtained from certain animal wastes such as cow dung and can be used locally. Research work has been going on to derive energy from municipal wastes and plastics also.

1.3   Modes of Combustion Processes

As mentioned earlier, combustion reaction occurs with heat release in all cases and along with emission of light of different intensities, in many cases. Modes of combustion can be defined based on the spatial extent where the oxidation reaction occurs. This can be either localized or spread over the entire combustion chamber.

When a fuel is mixed with the oxidizer in proper proportions, the resultant reactant mixture will be *flammable*, which means that when a sufficiently high-temperature ignition source is introduced, a combustion reaction will be initiated and sustained in the reactant mixture. When such a flammable reactant mixture is filled in a combustion chamber and a localized ignition source is instantaneously introduced, a combustion reaction involving the reactant mixture can be initiated. Based on the size and configuration of the combustion chamber, a *volumetric reaction* may occur over the entire chamber almost simultaneously. These types of volumetric reactions occur in small constant volume chambers such as bomb calorimeter, laboratory-scale constant pressure chambers, and well-stirred and plug flow reactors. The intensity of the reaction and the speed with which the chain reaction takes place depend upon the boundary conditions, the composition of the reactant mixture and pressure inside the combustion chamber. This can sometimes be so rapid that it is termed as an *explosion*. When the reaction chamber is longer (long tube containing reactant mixture) or larger (large spherical vessel containing reactant mixture), the localized ignition source will initiate the reaction only in a small region of the combustion chamber. This *localized reaction zone* is termed as the *flame*. This flame subsequently propagates through the combustion chamber, consuming the unburnt reactant mixture.

The propagation speed can be either subsonic, which is termed as *deflagration*, or supersonic, which is termed as *detonation*, based on certain conditions.

On the other hand, when the same reactant mixture is supplied continuously through a *burner* and ignition is initiated at the exit of the burner, a *stationary flame* (non-moving localized reaction zone) may be established near the burner exit, depending on the rate at which the reactant mixture is supplied. These modes of combustion of a flammable reactant mixture occurring without or with a flame are called *premixed combustion* processes. In these modes, the rates of the chemical reaction dictate the flame propagation. The flame propagation speed and its temperature are the important characteristics in a premixed combustion process.

When a fuel alone is supplied to the combustion chamber through one port and the oxidizer alone is supplied to the combustion chamber through another port, transport processes (such as diffusion and convection) occurring in the combustion chamber cause the fuel and the oxidizer to mix. When an ignition source is introduced *instantaneously* at an *appropriate* location, combustion reaction is initiated at locations where the fuel and the oxidizer have mixed in proper proportions. Therefore in such a process, a flame is formed or anchored at certain locations. This flame is called a *non-premixed flame*. Since fuel and oxidizer are supplied through different ports, locations where they are mixed in proper proportion and hence the flame anchoring location(s) will depend (among other things) upon the fuel and oxidizer flow rates. In these situations, the chemical reaction rate is usually very high and the transport processes, which control the mixing of fuel and oxidizer, govern the reaction zones. The extent of the flame, which is governed by the transport processes, is an important characteristic of non-premixed mode of combustion.

Both modes of combustion have their own advantages and shortcomings and are used in many practical applications based on the requirement. This requires a proper understanding of the underlying phenomena leading to a proper design of the combustion chamber/burner.

1.4 Emissions and Environment

Complete combustion of a hydrocarbon fuel results in all the carbon in the fuel being converted to CO_2 and all the hydrogen to water vapor. The overall oxidation reaction that accomplishes this takes place in several steps. Breakage of the atomic bonds in fuel and oxidizer species and the formation of intermediate fast reacting species called radicals happen first. The rapid reaction between the radicals and formation of final products happen subsequently. However, due to improper mixing, insufficient residence time and heat and radical losses due to transport processes, the reactions may not proceed to completion in a few regions in the combustion chamber. As a result of this, especially in the case of combustion of heavy fuels, *unburnt hydrocarbons*, CO and carbonaceous particles, called *soot*, are released. Even when the reaction is complete, if the temperature of the combustion chamber is high enough, species such as CO_2 and H_2O can dissociate to form CO and H_2

and these may be released to the atmosphere. Since atmospheric air is generally used as the oxidizer, under specific conditions of temperature and pressure, and when sufficient oxygen and residence time are available, nitrogen ceases to be inert and oxides of nitrogen (NO_x) are formed. While burning coal or gasoline, fuel-bound sulfur also takes part in the combustion reaction leading to the formation of oxides of sulfur (SO_x). While CO is toxic, unburnt hydrocarbons and soot are carcinogenic. In addition, the nanosized soot particles, when inhaled, can lead to respiratory illnesses. The oxides of nitrogen and sulfur, when released into the atmosphere, cause acid rain. This eventually destroys the fertility of the soil. Hence, emissions such as CO, unburnt hydrocarbons, soot, NO_x and SO_x are harmful to the environment and are appropriately classified as pollutants. Therefore, proper design of the burners and the combustion chamber, which will minimize the production of these emissions to the permissible levels prescribed by emission control boards, is essential.

Although CO_2 has been left out of the list of pollutants mentioned above, it must also be classified as such, since it is a greenhouse gas. In view of this, attempts are being made to develop mechanisms to capture CO_2. From an environment perspective, hydrogen alone may be termed as a clean fuel, as the main product of its combustion is only water vapor. Although combustion of H_2 with air may produce some amount of NO_x, this may be mitigated by proper design of the combustion chamber. Therefore, fuels with low C-to-H ratio, such as natural gas and synthetic gas, are desirable as their combustion produces a minimal amount of carbon-based emissions.

Review Questions

1. Define combustion.
2. What is a fossil fuel?
3. What are the pros and cons of a gaseous fuel?
4. What is calorific value?
5. How flash, fire and boiling points are defined for liquid fuels?
6. List the types of hydrocarbons present in crude oil.
7. How is a solid fuel classified?
8. List the advantages and disadvantages of liquid and solid fuels.

Chapter 2
Review of Combustion Thermodynamics and Kinetics

In this chapter, basic concepts in combustion stoichiometry, chemical thermodynamics and chemical kinetics as they relate to combustion are reviewed. *Stoichiometry* deals with the calculation of the required amount of oxidizer for a given amount of fuel. Heat and temperature calculations are carried out using the first law of thermodynamics. *Heat of reaction* (heat released during combustion) and *adiabatic flame temperature* (the maximum temperature attained) are important quantities required in the design of combustion chambers. Chemical *equilibrium* has its basis from the second law of thermodynamics. This study is useful in determining the composition of the products taking into account of dissociation of major species and leads to the estimation of realistic temperature of the products. As mentioned in the previous chapter, determining the residence time of the fuel–air mixture inside a combustor is important to ensure that the chemical reactions are completed within the combustion chamber. While the flow residence time can be estimated using fluid dynamics, estimation of chemical reaction time is carried out using *chemical kinetics*.

2.1 Combustion Stoichiometry

The first step in the design of a burner is the selection of the power rating. Once this is available, the required fuel flow rate may be calculated based on the calorific value of the chosen fuel. The next step is to calculate the amount of the oxidizer required. The methodology utilized for this purpose is called combustion stoichiometry. Since the temperatures involved in the combustion processes are normally very high, the density of the gas mixture will be considerably low and the gases are treated as ideal gases. Hence, the ideal gas equation of state is employed in this analysis.

Consider the following single-step reaction between hydrogen and oxygen:

© The Author(s), under exclusive license to Springer Nature Switzerland AG 2022
V. Raghavan, *Combustion Technology*,
https://doi.org/10.1007/978-3-030-74621-6_2

$$H_2 + \frac{1}{2}O_2 \rightarrow H_2O. \tag{2.1}$$

This reaction states that one mole or kilo-mole (kmol) of hydrogen reacts with half mole of oxygen to produce one mole of water vapor. One kmol of a species contains Avogadro number of molecules of that species. The number of moles of oxygen required is estimated by balancing of number of O atoms on both sides of the reaction. It can be noted that the total number of moles on the left-hand side (=1.5) is not same as that on the right-hand side (=1). Mass of a particular species is calculated by multiplying the number of moles with the molecular mass of the species. For instance, 1 kmol of hydrogen has a mass of 1×2 kg/kmol $= 2$ kg. Equation (2.1) may be written on a mass basis as,

$$2\,kg\,H_2 + 16\,kg\,O_2 \rightarrow 18\,kg\,H_2O. \tag{2.2}$$

It can be noted that the total mass of reactants ($2 + 16 = 18$ kg) is same as that of the product, and therefore, the mass is conserved automatically. Equation (2.2) also indicates that to burn 2 kg of hydrogen completely, 16 kg of oxygen is necessary. In other words, mass of oxygen required for the combustion of 1 kg hydrogen is 8 kg. This is usually represented as a ratio, namely, the mass of oxygen divided by the mass of fuel. In this case, the stoichiometric (theoretical) mass-based oxygen-to-hydrogen ratio is equal to 8.

In several applications, atmospheric air is used as the oxidizing species due to its abundant availability. In general, moisture and other impurities from the atmospheric air are removed using appropriate filters. Air is assumed to contain 21% oxygen by volume apart from nitrogen. The required number of moles of air for combustion is decided based on the required of number of moles of oxygen. Usually, air is represented as 1 kmol of oxygen and 3.76 kmol of nitrogen, so that the mole fractions of oxygen and nitrogen come out to be 0.21 [1/(1 + 3.76)] and 0.79 [3.76/(1 + 3.76)], respectively, which is the same as their volume fractions. Reaction (2.1), with air as the oxidizer, instead of O_2 may be is written as,

$$H_2 + \frac{1}{2}(O_2 + 3.76\,N_2) \rightarrow H_2O + \frac{3.76}{2}N_2. \tag{2.3}$$

In Eq. (2.3), the total number of moles of air is 0.5×4.76, since air has 1 mol e of oxygen and 3.76 moles of nitrogen. Mass of air for this case is calculated using molecular mass of air (MW_{air}) as,

$$0.5 \times 4.76 \times MW_{air}.$$

In turn, the molecular mass of air is calculated using the chain rule

$$0.21 \times MW_{oxygen} + 0.79 \times MW_{nitrogen} = 0.21 \times 32 + 0.79 \times 28 = 28.84 \text{ kg/kmol.}$$

Finally, the mass of air used in Eq. (2.3) is calculated as,

$$0.5 \times 4.76 \times 28.84 = 68.64 \ \text{kg}.$$

The ratio of mass of air to that of hydrogen comes out to be 34.32. The increase in mass-based oxidizer–fuel ratio, when air is used instead of oxygen, is apparent from this calculation. The amount of oxidizer calculated above is called the *stoichiometric* or *theoretical* air, which is required to completely burn 1 kmol of the fuel. The ratio is called *stoichiometric air–fuel ratio*. However, in general practice, the quantity of air supplied will be more than or, in some cases, less than this theoretical amount. The ratio of actual mass of the air supplied to the combustion system to the mass of the fuel is called the *actual* air–fuel ratio. The ratio of the stoichiometric air–fuel ratio to the actual air–fuel ratio is called *equivalence ratio*, represented by symbol ϕ. If ϕ is less than unity, the amount of air supplied is more than the theoretical air, if ϕ is equal to unity, the actual air is same as the theoretical air, and if ϕ is greater than unity, the actual air is less than the theoretical air. The equivalence ratio is also equal to the ratio of actual fuel–air ratio (ratio of mass of fuel to that of the actual mass of air) to the stoichiometric fuel–air ratio. The amount of air supplied is also represented as percentage theoretical air, calculated with respect to the theoretical amount of air.

For a general hydrocarbon fuel represented by C_xH_y, containing x carbon atoms and y hydrogen atoms, a general single-step reaction mechanism representing its complete combustion in air may be written as,

$$C_xH_y + a(O_2 + 3.76 \ N_2) \rightarrow xCO_2 + \frac{y}{2}H_2O + 3.76aN_2. \tag{2.4}$$

The value of "a" in Eq. (2.4) is obtained by balancing the number of carbon, hydrogen and oxygen atoms on both sides of the reaction and is given as, $a = x + y/4$. For instance, for methane (CH_4), $x = 1, y = 4$ and $a = 1 + 4/4 = 2$. The stoichiometric air–fuel ratio for methane combustion with air is thus $2 \times 4.76 \times 28.84/16 = 17.16$. Table 2.1 provides the stoichiometric air–fuel ratio for several hydrocarbons. In general for higher order straight chain hydrocarbons, the stoichiometric air–fuel ratio can be roughly taken as 15; that is, for one kg of fuel, approximately 15 kg of air is required theoretically.

Table 2.1 Stoichiometric air–fuel ratios of hydrocarbon combustion

Fuel	Stoichiometric air–fuel ratio
Ethane (C_2H_6)	16.02
Butane (C_4H_{10})	15.38
Decane ($C_{10}H_{22}$)	14.98
Ethylene (C_2H_4)	14.71
Benzene (C_6H_6)	13.20

When there is not enough oxygen available for complete oxidation, the hydrogen atom, being a highly reactive species, will first consume the required amount of oxygen for its oxidation to OH. The radical OH will contribute to the formation of H_2O through $H + OH \rightarrow H_2O$ and to the formation of CO_2 from CO through $CO + OH \rightarrow CO_2 + H$. However, the formation of water vapor is faster than that of CO_2. Therefore, not all the carbon in the fuel can oxidize completely to carbon dioxide, and some carbon monoxide remains without being oxidized. Furthermore, if temperature of the combustion chamber is quite high, around 2000 K, the carbon dioxide in the products will dissociate into carbon monoxide and oxygen. Water vapor, on the other hand, does not dissociate until the temperature reaches around 2800 K. Therefore, it is common to consider carbon monoxide in the products along with carbon dioxide. If the reactant mixture is fuel rich, oxygen will not be present in the products.

2.1.1 Analysis of Combustion Products

In order to determine whether complete combustion has taken place or not, the products of combustion must analyzed. In general, two methods are employed for this purpose, namely, the *gravimetric* analysis and the *volumetric* analysis. In the former, the masses of individual components in the gas mixture are obtained, and in the latter, the volumetric fractions of individual components are obtained. These analyses are carried out on a dry basis after removing the water vapor from the products. Based on the method employed, either mass fractions (Y_i) or volume fractions (V_{fi}) of all the components, except water vapor, in the product mixture are obtained. Since the gas mixture can be treated as an ideal gas, mole fractions (X_i) will be same as the volume fractions, as demonstrated by the Amagat's law of partial volumes. Using these data, and also performing the atom balances in the chemical reaction, the number of moles of water vapor produced may then be determined. In fact, with the product analysis, the C-to-H ratio may be estimated without even knowing the chemical formula of the fuel used.

The product mixture may be treated as an ideal gas by using a set of mixing rules. The molecular mass of the product mixture (MW_{mix}) is estimated using the relation,

$$MW_{mix} = \sum_i X_i MW_i = \frac{1}{\sum_i \frac{Y_i}{MW_i}} \tag{2.5}$$

The mole and mass fractions are related using the molecular mass of the mixture as,

$$X_i = Y_i \frac{MW_{mix}}{MW_i} \tag{2.6}$$

Further, the mass and the mole fractions obey the following identities.

$$\sum_i X_i = 1; \quad \sum_i Y_i = 1 \tag{2.7}$$

2.2 First Law Applied to Combustion

The amount of heat released in a combustion chamber due to a combustion reaction is known as the heat of reaction and is evaluated by applying the first law of thermodynamics to the combustor. The combustion process may occur in a constant pressure environment, as in diesel engines and gas turbines, or in a constant volume chamber, as in bomb calorimeter and gasoline engines.

The first law of thermodynamics applied to a system undergoing a process from state 1 to state 2, in which there are no changes in kinetic and potential energies, may be written as,

$$Q_{1-2} - W_{1-2} = \Delta U \tag{2.8}$$

For a constant pressure system, in the absence of work, other than the displacement work, Eq. (2.8) may be written as,

$$Q_{1-2} = \Delta U + W_{1-2} = U_2 - U_1 + p(V_2 - V_1) = H_2 - H_1 = H_P - H_R \tag{2.9}$$

That is, the heat transfer during the reaction process is calculated as the difference between the enthalpy of the product mixture and the enthalpy of the reactant mixture. This may be written as,

$$\Delta H_R = Q_{1-2} = H_P - H_R, \tag{2.10a}$$

where ΔH_R is called the heat of reaction. For combustion reaction taking place in a steady flow reactor, the heat of reaction may be calculated as,

$$\Delta H_R = \dot{Q} = \dot{m}(h_P - h_R), \tag{2.10b}$$

where $\dot{m}$ is mass flow rate of reactant into the reactor. As stated in Chap. 1, for an exothermic reaction, the enthalpy of the products is less than that of the reactants, so that heat is evolved during the combustion reaction. Therefore, the heat of reaction, calculated using Eq. (2.10a, b), is negative. This will be the heat gained by the surroundings and is called the heat of combustion, ΔH_c. Heat of combustion has the same magnitude as the heat of reaction but with a negative sign. For a constant volume system, the displacement work is zero, and Eq. (2.8) may be written as,

$$Q_{1-2} = \Delta U = U_2 - U_1$$

That is, the heat transfer during the reaction process is the difference between the internal energy of the product mixture and the internal energy of the reactant mixture. This may be written in terms of enthalpies as,

$$\Delta H_R = Q_{1-2} = U_P - U_R = \left(H_p - n_p R_u T_p\right) - (H_R - n_R R_u T_R) \qquad (2.11)$$

In Eq. (2.11), n_P and n_R are the numbers of moles of reactant and product mixtures, respectively, T_P and T_R are the temperatures of reactant (initial state) and product (final state), respectively, and R_u is universal gas constant (8314.15 J/kmol K). In this case, the heat of combustion has the same magnitude as the difference in the internal energy of the product and that of the reactant, but with a negative sign.

2.2.1 Standard Enthalpy

In order to evaluate the enthalpies of the reactant and product mixtures in Eqs. (2.10a, b) and (2.11), the absolute enthalpies of the individual species that constitute the mixtures must be known. The absolute enthalpy of a species i is expressed in molar basis as,

$$\bar{h}_i(T) = \bar{h}_{f,i}^o(T_{\text{ref}}) + \Delta h_i(T) \qquad (2.12)$$

In Eq. (2.12), the first term in the right-hand side, $\bar{h}_{f,i}^o(T_{\text{ref}})$, is called the standard enthalpy of formation for the given species. It is calculated at the standard-state temperature (298 K) and at atmospheric pressure. It is defined as the increase in enthalpy, when one mole of a compound is formed at constant pressure from its natural elements, which are in the standard state (1 atmosphere, 298 K) and the compound, after its formation, is also brought to the standard state. For naturally occurring elements such as O_2, H_2, N_2 and C (solid), the standard enthalpy of formation is assigned a value of zero.

The enthalpy of formation of a compound may be determined from the energies associated with the bonds that make up the molecular structure of the compound or through experiments. Energy is absorbed when bonds break, and energy is released when the bonds are formed. The enthalpy of formation of a compound can be calculated by subtracting the bond energies for bonds that are formed from the total bond energy for the bonds that are broken during the formation of the compound from its constitutive elements. This is illustrated through the following example.

Consider formation of water from its basic elements at 298 K, 1 atmosphere, given by the single-step reaction, $2H_2 + O_2 \rightarrow 2H_2O$. Two hydrogen atoms are bonded by a single bond in hydrogen molecule. Two oxygen atoms are bonded by a double bond in oxygen molecule. The energy associated with a single bond will be less than that of a double bond. A water molecule contains two H atoms bonded to an O atom by two single bonds. That is, a water molecule has two O–H bonds. The energy required

Fig. 2.1 Calculation of enthalpy of formation using bond energies

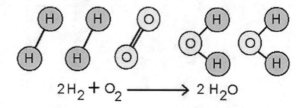

$$2H_2 + O_2 \longrightarrow 2H_2O$$

for the formation of four O–H bonds minus the energy needed for the destruction of two H–H and one O=O bond will give the enthalpy of formation of two water molecules. This is schematically shown in Fig. 2.1. It is calculated as follows:

$$4 \times \text{Energy}\,(O - H) - 2 \times \text{Energy}\,(H - H) - \text{Energy}\,(O = O)$$
$$= \text{Enthalpy of formation of two } H_2O \text{ molecules}$$

Energy absorbed when a H–H bond breaks $= 432$ kJ/mol.
Energy absorbed when a O=O bond breaks $= 494$ kJ/mol.
Net energy associated with bonds being broken $= 2 \times 432 + 492 = 1358$ kJ.
Energy released when a O–H bond forms $= 459$ kJ/mol.
Enthalpy of formation of two moles of $H_2O = 1358 - 4 \times 459 = -478$ kJ.
Enthalpy of formation per mol of water vapor $= -239$ kJ/mol of H_2O.
The second term in the right-hand side of Eq. (2.12) is the enthalpy associated with the temperature change alone, and is called the sensible enthalpy. This is the increase in the enthalpy due to an increase in the temperature from the standard value (298 K) to a higher temperature, $T(K)$. For ideal gases, which are also thermally perfect, enthalpy and internal energies are functions of temperature alone. The sensible enthalpy is calculated by integrating the specific heat at constant pressure, which is generally expressed as a polynomial in temperature. That is,

$$\Delta h_i(T) = \int_{298}^{T} c_{p,i} dT$$

The absolute enthalpies of O atom and O_2 molecule as a function of temperature are shown in Fig. 2.2. It should be noted that at 298 K, the absolute enthalpy of O_2 is zero, because it is naturally occurring and that of O atom is nonzero.

2.2.2 Standard Heat of Reaction

The standard heat of reaction is the heat released when the reactants at 298 K and 1 atmosphere pressure react to form products, and the products are cooled to 298 K. This is also referred to as calorific value or heating value, expressed as J per kg or

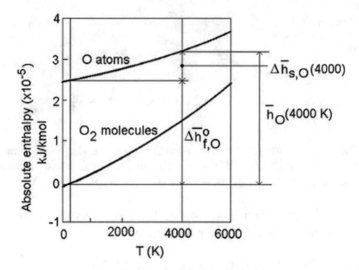

Fig. 2.2 Variation of absolute enthalpies of O atom and O_2 molecule

per kmol of the fuel. Further, the heating value is termed as higher heating value when all the water vapor in the products are condensed to liquid form; consequently, the latent heat of vaporization of water is also added to the heat of combustion. The lower heating value is realized when the water vapor in the products is not allowed to condense at all. When higher order hydrocarbon fuels are used, they exist in liquid phase at the standard state. Therefore, while calculating the heat of reaction, the energy needed for phase change has to be taken into account. For instance, n-Decane, which exists in liquid form at the standard state, can have four possible heat of combustion values based on whether the fuel and water are in liquid or vapor phases. This is illustrated in Fig. 2.3.

In Fig. 2.3, subscripts g and l represent the gas and liquid phases, respectively, and $\Delta h_{c,g-l}$ represents the higher heating value.

Example 2.1 Determine the standard heat of reaction of methane burning in air, in a constant pressure environment, assuming complete combustion takes place.

Solution
The chemical reaction for complete combustion of methane in air is written as,

$$CH_4 + 2(O_2 + 3.76\,N_2) \rightarrow CO_2 + 2H_2O + 7.52\,N_2. \qquad (2.13)$$

$$\text{Standard heat of reaction} = H_{\text{products}} - H_{\text{reactants}}$$
$$= \left(\bar{h}_{CO_2} + 2\bar{h}_{H_2O} + 7.52\bar{h}_{N_2}\right) - \left(\bar{h}_{CH_4} + 2\bar{h}_{O_2} + 7.52\bar{h}_{N_2}\right),$$

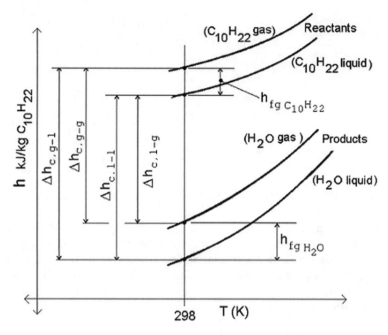

Fig. 2.3 Heating values of decane when the fuel and water are in different phases

where the over-bar indicates a molar basis (J/kmol). Since both reactants and products have to be at the standard-state temperature to evaluate the standard heat of reaction, only enthalpies of formation contribute to the absolute enthalpies. It should be noted that if the products are at a temperature higher than 298 K, then sensible enthalpies would be nonzero. For several species, the enthalpies of formation, sensible enthalpies and specific heat values as a function of temperature are available in NIST Web site (http://webbook.nist.gov/chemistry/). Substituting the enthalpies of formation of CH_4, CO_2 and H_2O, the standard heat of reaction of methane–air stoichiometric reaction per kmol of fuel is obtained as,

$$\Delta H_R = [(-393.52) + 2(-241.83) + 7.52 \times 0] - [-74.5 + 2 \times 0 + 7.52 \times 0]$$
$$= -802.68 \, \text{kJ/mol} - CH_4$$
$$= -802.68 \times 10^3 \, \text{kJ/kmol} - CH_4$$

Dividing this by molecular mass of methane, the standard heat of reaction per kg of fuel is obtained as $-50,167.5$ kJ/kg-CH_4. This value is lower heating value as the water in the products is in vapor state. It is known that for one kmol of methane, 2 kmol of oxygen is required. Thus, based on kmol of oxygen used, the standard heat of reaction may be estimated as,

$$\Delta H_{R,O_2} = -802.68 \times 10^3/2 = -401.34 \times 10^3 \, \text{kJ/kmol} - O_2.$$

Table 2.2 Standard heat of combustion for several fuels

Fuel	Δh_c MJ per kg-fuel	Δh_c MJ per kg-O_2
Methane	50.2	12.6
Propane	46.1	12.7
Butane	45.5	12.7
Methanol	19.9	13.3
Ethanol	27.6	13.2
Benzene	40.0	13.0
n-heptane	44.7	12.7

Using the molecular mass of oxygen, the standard heat of reaction per kg of oxygen may be estimated as $-12{,}542$ kJ/kg-O_2.

The standard heat of combustion, Δh_c, is same in magnitude as ΔH_R, but with an opposite sign. For several fuels, Δh_c values per kg- fuel as well as per kg oxygen have been reported in Table 2.2. It is interesting to note that the heat of combustion per kg oxygen is almost constant with a value of 13.0 ± 0.4 MJ/kg-O_2. Therefore, if oxygen consumption is measured carefully, the heat released during combustion can be estimated, irrespective of the fuel used.

2.2.3 Adiabatic Flame Temperature

After the completion of combustion inside a combustion chamber, the temperature of the products will be a maximum when all the heat released during the combustion is absorbed by the products, i.e., when the heat transfer to the surroundings is zero. For a constant pressure combustion occurring in gas turbines or in a flow process, if combustion occurs in an adiabatic environment, the absolute enthalpy of the reactants at the initial state (T_i, p) will be equal to the absolute enthalpy of the products at the final state (T_{ad}, p). That is, the enthalpy of the reactants is same as that of the products:

$$H_{reac}(T_i, p) = H_{prod}(T_{ad}, p). \tag{2.14a}$$

This is illustrated in Fig. 2.4.

On the other hand, for constant volume combustion, such as in a gasoline engine, if the fuel–air mixture burns adiabatically, the absolute internal energy of the reactants at the initial state (T_i, p_i) will be equal to the absolute internal energy of the products at the final state (T_{ad}, p_f). That is,

$$U_{reac}(T_i, p_i) = U_{prod}(T_{ad}, p_f). \tag{2.14b}$$

This may be rewritten as,

Fig. 2.4 Illustration of adiabatic flame temperature for constant pressure combustion

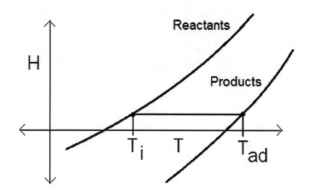

$$H_{\text{reac}}\left(T_i, p_i\right) - H_{\text{prod}}\left(T_{\text{ad}}, p_f\right) - V\left(p_i - p_f\right) = 0. \tag{2.14c}$$

If the product composition is known, either Eq. (2.14a) or Eq. (2.14b) can be employed to estimate the adiabatic temperature. This is illustrated using an example.

Example. 2.2 Determine the adiabatic flame temperature during methane combustion in air in a constant pressure environment, assuming that complete combustion takes place. The reactant temperature may be taken as 298 K.

Solution

For constant pressure combustion process,

$$H_{\text{reac}}\left(T_i, p\right) - H_{\text{prod}}\left(T_{\text{ad}}, p\right) = 0.$$
$$\Rightarrow \bar{h}_{\text{CH}_4}(T_i) + 2\bar{h}_{\text{O}_2}(T_i) + 7.52\bar{h}_{\text{N}_2}(T_i)$$
$$- \bar{h}_{\text{CO}_2}(T_{\text{ad}}) - 2\bar{h}_{\text{H}_2\text{O}}(T_{\text{ad}}) - 7.52\bar{h}_{\text{N}_2}(T_{\text{ad}}) = 0 \tag{2.15a}$$

$$\Rightarrow \bar{h}^o_{f,\text{CH}_4} - \bar{h}^o_{f,\text{CO}_2} - \Delta\bar{h}_{\text{CO}_2}\big|^{T_{\text{ad}}}_{298} - 2\bar{h}^o_{f,\text{H}_2\text{O}} - 2\Delta\bar{h}_{\text{H}_2\text{O}}\big|^{T_{\text{ad}}}_{298} - 7.52\Delta\bar{h}_{\text{N}_2}\big|^{T_{\text{ad}}}_{298} = 0 \tag{2.15b}$$

or

$$\bar{h}^o_{f,\text{CH}_4} - \bar{h}^o_{f,\text{CO}_2} - \int^{T_{\text{ad}}}_{298} \bar{c}_{p,\text{CO}_2}dT - 2\bar{h}^o_{f,\text{H}_2\text{O}} - 2\int^{T_{\text{ad}}}_{298} \bar{c}_{p,\text{H}_2\text{O}}dT - 7.52\int^{T_{\text{ad}}}_{298} \bar{c}_{p,\text{N}_2}dT = 0 \tag{2.15c}$$

Here, $\bar{c}_{p,i}$ represents molar specific heat at constant pressure and $\Delta\bar{h}$ is the sensible enthalpy. As mentioned earlier, lookup tables are available for sensible enthalpy of different species as a function of temperature. By assuming a value of T_{ad}, the sensible enthalpy values of CO_2, H_2O and N_2, at that temperature, should be looked up. Then, using those values, Eq. (2.15c) should be verified, if it approaches a value of zero within a given tolerance. The value of T_{ad} should be iterated such that Eq. (2.15c) is ultimately satisfied. Alternatively, temperature polynomials are reported in NIST to evaluate the specific heat of a given species as a function of temperature. By using such polynomials for CO_2, H_2O and N_2, and integrating them as expressed in Eq. (2.15c), the adiabatic flame temperature, T_{ad}, can be estimated. It should be noted that the resulting equation for T_{ad} is non-linear and should be solved using techniques such as Newton–Raphson method.

The adiabatic flame temperature for stoichiometric reaction of methane in air is evaluated as 2325 K by solving a nonlinear equation for T_{ad} as in Eq. (2.15b). This value can also be obtained when Eq. (2.15c) is used. Initially, T_{ad} is guessed as 2200 K. Sensible enthalpy values of species at this temperature are used in Eq. (2.15c), which results in a positive residue equal to 56.3. When a value of 2400 K is used for T_{ad} and substituting the sensible enthalpy values looked up at this temperature, a negative residue of -31.8 results. Therefore, by interpolation, a value of 2328 K is obtained for T_{ad}.

In the case of constant volume combustion, if the composition of the products is known, the first two terms in Eq. (2.14c) are evaluated as illustrated in Example 2.2. This is because the gas mixture is treated as an ideal gas, and hence, its enthalpy does not depend upon pressure. The third term is evaluated using the equation of state; that is,

$$V\left(p_i - p_f\right) = n_R R_u T_i - n_P R_u T_{ad}$$

On the other hand, if the product composition is not known, then the procedure becomes complex. First law alone cannot be employed to arrive at the adiabatic flame temperature. Based on the pressure and temperature of the combustion products, major product species such as CO_2 and H_2O may dissociate giving rise to species such as CO, H_2 and so on. When chemical equilibrium prevails, the extent of dissociation and the product composition are determined using the concepts of the second law of thermodynamics.

2.3 Second Law Applied to Combustion

In a combustion chamber, after a chemical reaction is almost complete, the temperature may reach high values of the order of above 2000 K. Under this condition, major products such as CO_2 and H_2O, and excess oxidizer species such as N_2 and O_2 if any, dissociate to produce species such as CO, H_2, OH, H, O, N and NO and so on. The mass fractions of these species in the product mixture depend on the temperature and pressure inside the combustion chamber.

Consider dissociation of CO_2 occurring in an isolated system, given by the reaction,

$$CO_2 \rightarrow CO + 1/2O_2 \tag{2.16}$$

Based on the temperature and pressure inside the combustion chamber, if out of one mole of CO_2, if α moles ($\alpha < 1$) dissociate, then $(1 - \alpha)$ moles of CO_2, α moles of CO and $\alpha/2$ moles of O_2 will be present in the product mixture. As the reaction proceeds, at a given value of α, say α_e, changes in the number of moles of these species are not seen for the given temperature, pressure and volume of the system. For the isolated system, this represents an *equilibrium state* and the value of α_e will determine the equilibrium composition. To estimate the value of α_e, concepts from second law of thermodynamics are used.

From the second law of thermodynamics, it is known that *entropy* (S) of a system changes due to heat transfer; it increases when heat is transferred to the system or it decreases if the heat is rejected by the system. Also, entropy increases due to irreversibility. Therefore, for an isolated system, the entropy may either remain constant or increase. Constant entropy state is called the equilibrium state. When a constant volume process is in equilibrium, if entropy is specified to be constant, the internal energy (U) assumes the minimum value. On the other hand, if internal energy is specified to be a constant, then entropy assumes a maximum value. If neither S nor U is constant, then for the constant volume system at equilibrium, both S and U together adjust such that the quantity $(U - TS)$ assumes a minimum value. This quantity is a derived property called *Helmholtz free energy* (A),

$$A = U - TS, \text{ or } a = u - Ts$$

where a, u and s are specific properties. Following the definition of enthalpy, defined using a constant pressure process, $(H = U + pV)$, another free energy, called *Gibbs free energy* (G), may be defined as,

$$G = A + pV = U + pV - TS = H - TS, \text{ or } g = h - Ts.$$

Here, g and h are specific properties. Similar to the constant volume process, where Helmholtz free energy attains a minimum value at equilibrium, in a constant pressure process, the Gibbs free energy attains a minimum value at equilibrium. For instance, if the mixture entropy and mixture Gibbs free energy are plotted in Fig. 2.5 against the extent of dissociation of CO_2, $(1 - \alpha)$, as considered in Eq. (2.16), based on temperature and pressure of the mixture. It can be seen that entropy attains a local maximum and Gibbs free energy attains a local minimum at the equilibrium value of $\alpha = \alpha_e$. The condition for equilibrium is derived as follows. Consider a constant volume process. Any change in A may be expressed as,

$$dA = dU - TdS - SdT$$

From the definition of entropy change and I law, $TdS = dU + pdV$, and so,

$$dA = -pdV - SdT.$$

Fig. 2.5 Variation of
entropy and Gibbs free
energy of the mixture with
extent of reaction

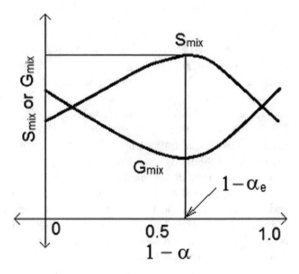

It is clear that A is a function of V and T with $-p$ being the potential function associated with the volume and $-S$ that associated with the temperature. Consider a binary mixture, having n_1 moles of component 1 and n_2 moles of component 2. Due to chemical reaction, the number of moles of these components in the mixture may change. As a result, the properties, including A, may change. Therefore, for a reacting system, A is a function of V, T, n_1 and n_2, and the change in Helmholtz free energy may then be expressed as,

$$dA = -pdV - SdT + \mu_1 dn_1 + \mu_2 dn_2.$$

Here, μ_1 and μ_2 are the potential functions associated with n_1 and n_2, respectively. These are called *chemical potentials*. This may be extended to mixture containing N species. Change in A may further be written as,

$$dA = \left(\frac{\partial A}{\partial V}\right)_{T,n_1,n_2} dV + \left(\frac{\partial A}{\partial T}\right)_{V,n_1,n_2} dT + \left(\frac{\partial A}{\partial n_1}\right)_{V,T,n_2} dn_1 + \left(\frac{\partial A}{\partial n_2}\right)_{V,T,n_1} dn_2.$$

It follows that,

$$-p = \left(\frac{\partial A}{\partial V}\right)_{T,n_1,n_2} ; -S = \left(\frac{\partial A}{\partial T}\right)_{V,n_1,n_2} ; \mu_1 = \left(\frac{\partial A}{\partial n_1}\right)_{V,T,n_2} ; \mu_2 = \left(\frac{\partial A}{\partial n_2}\right)_{V,T,n_1}.$$

From these definitions, μ_1 and μ_2 are the *partial molal Helmholtz functions*, expressed as $\bar{a}_1$ and $\bar{a}_2$, which are defined at constant volume and temperature. Therefore, change in Helmholtz free energy may be written as,

$$dA = -pdV - SdT + \bar{a}_1 dn_1 + \bar{a}_2 dn_2.$$

At constant volume and constant temperature,

$$dA = \bar{a}_1 dn_1 + \bar{a}_2 dn_2.$$

Further, at the equilibrium state, there is no change in the number of moles of any species. Therefore, the condition for equilibrium for a constant volume process is given by $dA = 0$.

Similarly, for a constant pressure process, the above procedure may be adopted using Gibbs free energy as demonstrated below:

$$dG = dH - TdS - SdT.$$

From the definition of enthalpy and entropy change, and I law, $TdS = dH - Vdp$, and so,

$$dG = -Vdp - SdT.$$

Thus, G is a function of p and T with $-V$ being the potential function associated with the pressure and $-S$ that associated with the temperature. For a binary reacting system, G is a function of p, T, n_1 and n_2 and the change in Gibbs free energy may then be expressed as,

$$dG = -Vdp - SdT + \mu_1 dn_1 + \mu_2 dn_2.$$

Change in G can further be written as,

$$dG = \left(\frac{\partial G}{\partial p}\right)_{T,n_1,n_2} dp + \left(\frac{\partial G}{\partial T}\right)_{p,n_1,n_2} dT + \left(\frac{\partial G}{\partial n_1}\right)_{p,T,n_2} dn_1 + \left(\frac{\partial G}{\partial n_2}\right)_{p,T,n_1} dn_2.$$

It follows that,

$$-V = \left(\frac{\partial G}{\partial p}\right)_{T,n_1,n_2} ; -S = \left(\frac{\partial G}{\partial T}\right)_{p,n_1,n_2} ; \mu_1 = \left(\frac{\partial G}{\partial n_1}\right)_{p,T,n_2} ; \mu_2 = \left(\frac{\partial G}{\partial n_2}\right)_{p,T,n_1}.$$

Here, μ_1 and μ_2 are equal to the *partial molal Gibbs functions*, expressed as $\bar{g}_1$ and $\bar{g}_2$, which are defined at constant pressure and temperature. Therefore, change in Gibbs free energy may be written as,

$$dG = -Vdp - SdT + \bar{g}_1 dn_1 + \bar{g}_2 dn_2.$$

At constant pressure and constant temperature,

$$dG = \bar{g}_1 dn_1 + \bar{g}_2 dn_2.$$

Therefore, the condition for equilibrium for constant pressure process, when the number of moles of all the species are constant, is given by $dG = 0$.

The partial molal Gibbs function may be estimated at any temperature and pressure as,

$$\bar{g}(T, p) = \bar{h}(T, p) - T\,\bar{s}(T, p).$$

Since all the species are assumed to behave as ideal gases, enthalpy of the species is independent of pressure and evaluated using Eq. (2.12). Entropy of any species, which is an ideal gas, may be expressed as,

$$\bar{s}_i(T, p_i) = \bar{s}_i(T_{ref}, p_0) + \int_{T_{ref}}^{T} \bar{c}_{p,i}\frac{dT}{T} - R_u \ln\frac{p_i}{p_0} = \bar{s}_i(T, p_0) - R_u \ln\frac{p_i}{p_0}.$$

Here, p_0 is the atmospheric pressure and p_i is partial pressure of ith species. Using this,

$$\bar{g}_i(T, p_i) = \bar{h}_i(T) - T \times \bar{s}_i(T, p_0) - R_u T \ln\frac{p_i}{p_0} = \bar{g}_i(T, p_0) - R_u T \ln\frac{p_i}{p_0}.$$
$$(2.17)$$

Polynomials for evaluating entropy of species as a function of temperature are available in the NIST Web site. Using the enthalpy and entropy data, Gibbs free energy of the species may be estimated using Eq. (2.17). For a constant volume process, partial molal Helmholtz function of a species may be calculated from the partial molal Gibbs function as,

$$\bar{a}_i(T, p_i) = \bar{g}_i(T, p_i) - p_i V.$$

Using the partial molal Gibbs function of individual species, Gibbs free energy of a mixture of ideal gas may be expressed as,

$$G_{mix} = \sum N_i \bar{g}_i(T, p) = \sum N_i \left\{ \bar{g}_i(T, p_0) + R_u T \ln\left(\frac{p_i}{p_0}\right) \right\}. \qquad (2.18)$$

Here, N_i is the number of moles of ith species. For fixed T and p, the equilibrium condition is given as $dG_{mix} = 0$. This implies,

$$\sum_i dN_i \left(\bar{g}_i(T, p_0) + R_u T \ln\left(\frac{p_i}{p_0}\right) \right) + \sum_i N_i d\left(\bar{g}_i(T, p_0) + R_u T \ln\left(\frac{p_i}{p_0}\right) \right) = 0.$$
$$(2.19a)$$

The second term in Eq. (2.19a) is zero, since, as the pressure is constant, $d[\ln(p_i)]$ $= dp_i/p_i = 0$. Therefore, the condition for equilibrium may be written as,

$$\sum_i dN_i \left(\bar{g}_i(T, p_0) + R_u T \ln\left(\frac{p_i}{p_0}\right) \right) = 0. \tag{2.19b}$$

The condition for chemical equilibrium for a constant volume process may be estimated in terms of partial molal Helmholtz function by following a similar procedure.

2.3.1 Equilibrium Constant

Consider a general reaction taking place under a constant pressure process,

$$aA + bB \leftrightarrow cC + dD. \tag{2.20}$$

Here, A and B are reactants, C and D are products, for the forward reaction, and $a - d$ are the stoichiometric coefficients. The change in the number of moles of a species is directly proportional to the corresponding stoichiometric coefficient, and hence,

$$dN_A = -\kappa a; dN_B = -\kappa b; dN_C = +\kappa c \text{ and } dN_D = +\kappa d.$$

By substituting this in Eq. (2.19), the constant of proportionality κ will cancel out and the equilibrium condition may be written as,

$$-a\left[\bar{g}_A(T, p_0) + R_u T \ln\left(\frac{p_A}{p_0}\right)\right] - b\left[\bar{g}_B(T, p_0) + R_u T \ln\left(\frac{p_B}{p_0}\right)\right] =$$
$$-c\left[\bar{g}_C(T, p_0) + R_u T \ln\left(\frac{p_C}{p_0}\right)\right] - d\left[\bar{g}_D(T, p_0) + R_u T \ln\left(\frac{p_D}{p_0}\right)\right]$$

By grouping the partial molal Gibbs function on one side, the condition for equilibrium may be written as,

$$-[c \times \bar{g}_C(T, p_0) + d \times \bar{g}_D(T, p_0) - a \times \bar{g}_A(T, p_0) - b \times \bar{g}_B(T, p_0)] =$$
$$= R_u T \ln\left[\frac{(p_C/p_0)^c (p_D/p_0)^d}{(p_A/p_0)^a (p_B/p_0)^b}\right] \tag{2.21}$$

The expression in the left-hand side of Eq. (2.21) is defined as the standard state Gibbs function change $\Delta G(T, p_0)$, given by,

$$\Delta G(T, p_0) = c \times \bar{g}_C(T, p_0) + d \times \bar{g}_D(T, p_0) - a \times \bar{g}_A(T, p_0) - b \times \bar{g}_B(T, p_0). \tag{2.22}$$

The equilibrium constant based on partial pressures, K_p, is defined as,

$$K_p = \left[\frac{(p_C/p_0)^c (p_D/p_0)^d}{(p_A/p_0)^a (p_B/p_0)^b} \right] \tag{2.23}$$

Finally, the condition for equilibrium for constant pressure process for a general system can be written as,

$$\Delta G(T, p_0) = -R_u T \ln K_p, \tag{2.24}$$

or

$$K_p = \exp\left[\frac{-\Delta G(T, p_0)}{R_u T} \right] \tag{2.25a}$$

In terms of standard enthalpy and entropy changes, equilibrium constant may be expressed as,

$$K_p = \exp\left[\frac{-\Delta H(T)}{R_u T} \right] \exp\left[\frac{-\Delta S(T, p_0)}{R_u} \right] \tag{2.25b}$$

2.3.2 Determination of Equilibrium Composition

It is clear from Eq. (2.22) that the standard Gibbs function change is a function of temperature only. Therefore, from Eq. (2.25a), K_p is also a function of temperature only. By estimating the partial molal Gibbs function of the species considered, the Gibbs function change may be estimated and from that K_p may be determined. Then, Eq. (2.23) is used. The ratio p_i/p_0 is called activity coefficient. If the combustion takes place at a pressure p, then using the mole fraction of ith species, X_i, the activity coefficient may be written as $X_i p/p_0$. For instance, consider the general reaction given in Eq. (2.20). For this reaction, K_p is expressed in terms of mole fractions as,

$$K_p = \frac{X_C^c X_D^d}{X_A^a X_B^b} \left(\frac{p}{p^0} \right)^{c+d-a-b}. \tag{2.26}$$

Since K_p is already known from Eq. (2.25a), the mole fractions of the products for the reaction taking place at the pressure p may be calculated after some algebra. In fact, the mole fractions may be expressed using a single variable, α, which is the extent of reaction as given in Eq. (2.16). For this reaction, equilibrium value of α, which is α_e, is calculated as follows:

At a given temperature, standard Gibbs function change is calculated as,

$$\Delta G(T, p_0) = \bar{g}_{CO}(T, p_0) + 0.5 \times \bar{g}_{O_2}(T, p_0) - \bar{g}_{CO_2}(T, p_0)$$

The equilibrium constant is calculated as,

$$K_p = \exp\left[\frac{-\Delta G(T, p_0)}{R_u T}\right]$$

In terms of partial pressures, K_p is expressed as,

$$K_p = \frac{X_{CO} X_{O_2}^{0.5}}{X_{CO_2}}\left(\frac{p}{p^0}\right)^{0.5} = \frac{\left(\frac{\alpha}{1+\alpha/2}\right)\left(\frac{\alpha/2}{1+\alpha/2}\right)^{0.5}}{\left(\frac{1-\alpha}{1+\alpha/2}\right)}\left(\frac{p}{p^0}\right)^{0.5}$$

This results in a nonlinear equation for α. After solving this equation and considering the physically possible roots, the value of α_e may be determined. With respect to the effect of pressure, it has been observed that when pressure increases at constant temperature, the backward reaction will be favored. Dissociation normally occurs at high temperatures at atmospheric pressure.

2.3.3   Equilibrium Flame Temperature

The adiabatic flame temperature calculated considering the equilibrium products of combustion is called equilibrium flame temperature. This has to be determined iteratively, using a procedure similar to the one used earlier for the adiabatic flame temperature.

a.   Assume an adiabatic flame temperature, T_f,
b.   Evaluate K_p, from standard Gibbs function change calculated for the equilibrium reaction, using Eqs. (2.22 and 2.25a, b).
c.   Express K_p in terms of mole fractions as in Eq. (2.26) and solve for the number of moles in the products using a non-linear algebraic equation solver,
d.   Find the enthalpy of the product mixture, H_P (T_f) and compare it with the enthalpy of the reactants, H_R (T_i).
e.   If $H_P = H_R$, then T_f is the correct adiabatic flame temperature considering equilibrium products, else, iterate by changing the value of T_f and repeat steps ($b - e$) again.

Consider Eq. (2.13), which represents the stoichiometric combustion of methane in air. The adiabatic flame temperature for this reaction, considering stable products, has been calculated in Example 2.2. If CO_2 dissociates, as expected, products will contain CO as well. The reaction including CO in the products may be written as,

$$CH_4 + 2(O_2 + 3.76N_2) \rightarrow aCO_2 + bCO + cO_2 + 2H_2O + 7.52N_2.$$

To find the values of $a - c$, atom balances and equilibrium constant for dissociation reaction of CO_2 (Eq. 2.16) are used.

C balance: $1 = a + b$

O balance: $4 = 2a + b + 2c$

$$K_p = \frac{\left(\frac{b}{a+b+c}\right)\left(\frac{c}{a+b+c}\right)^{0.5}}{\left(\frac{a}{a+b+c}\right)} \left(\frac{p}{p^0}\right)^{0.5} = \frac{b \cdot c^{0.5}}{a(a+b+c)^{0.5}} \left(\frac{p}{p^0}\right)^{0.5}$$

At a given temperature, value of K_p may be calculated, and using the above three equations, values of a to c may be determined. The iterative procedure outlined is then used to obtain the adiabatic flame temperature. For methane–air combustion at atmospheric pressure, considering the dissociation of CO_2, the adiabatic flame temperature is obtained as 2232 K.

2.4 Chemical Kinetics

The knowledge of the rate at which a combustion reaction takes place is essential in order to ensure that the reactants stay inside the combustion chamber for a sufficiently long enough period of time, known as the residence time, within which the combustion reaction would be completed. Chemical reactions occur when the molecules of one of the reacting species collide with the molecules of another reacting species with proper orientation and energy to produce new molecules of product species. The chemical reaction essentially involves breaking of atomic bonds in the reactant species molecules and forming of new atomic bonds to produce molecules of the product species. Chemical bonds are broken during an impact or collision that occurs with sufficient energy. Energy content within an atomic bond depends on the nature of the atoms and on geometrical factors. Thus, the energy content of the products of collision may be different from the energy content of the colliding (reacting) molecules. In general, energy should be supplied to break a bond. Depending upon the pressure and temperature, new atomic bonds of the product species are formed, and in this process, energy is released. If energy released is greater than energy supplied, then such a reaction is called an exothermic reaction.

2.4.1 Types of Reactions

Combustion reactions may be classified as global reactions and elementary reactions. A global reaction is usually written in a single step. For example, for the complete combustion of methane in air, the *global reaction* may be written as,

$$CH_4 + 2(O_2 + 3.76 N_2) \rightarrow CO_2 + 2H_2O + 7.52 N_2.$$

In writing this global reaction, it has been assumed that nitrogen is considered as an inert species and the major product species CO_2 and H_2O are stable.

However, at high temperatures dissociation of the major product species is inevitable. As mentioned earlier, based on the temperature and pressure, CO_2 may dissociate to form CO and O. Similarly, H_2O may dissociate to form H_2, H and OH. In order to predict the heat of the reaction and the flame temperature with reasonable accuracy, in addition to the major species, several minor species should be included in the product mixture. The combustion of a general hydrocarbon, C_xH_y, in air may be described by the following global single-step mechanism, including several minor species in the product mixture,

$$C_xH_y + a(O_2 + 3.76 N_2) \rightarrow bCO_2 + cCO + dH_2O + eH_2 + fO_2$$
$$+ gN_2 + hH + kO + mOH + nNO + oN \qquad (2.27)$$

At a given temperature and pressure, the equilibrium composition of the product mixture may be determined, if the number of moles of each product species (coefficients $b - o$) are known. Eleven equations are required in order to determine these eleven coefficients. From the atom balances of C, H, N and O, four equations can be formed. For constructing seven more equations, seven chemical reactions involving the species present in the product mixture have to be considered. If the equilibrium constants of these reactions are known at the given temperature, then the equilibrium composition may be determined at the given pressure. For the above example, the following seven reactions may be considered in order to evaluate the coefficients, $b - o$.

$$CO + H_2O \Leftrightarrow CO_2 + H_2 \qquad (2.28a)$$

$$CO + 1/2O_2 \Leftrightarrow CO_2 \qquad (2.28b)$$

$$H_2O \Leftrightarrow OH + 1/2H_2 \qquad (2.28c)$$

$$O + N_2 \Leftrightarrow NO + N \qquad (2.28d)$$

$$H_2 \Leftrightarrow 2H \qquad (2.28e)$$

$$O_2 \Leftrightarrow 2O \qquad\qquad (2.28f)$$

$$N_2 \Leftrightarrow 2N \qquad\qquad (2.28g)$$

The equilibrium constants of the reactions in (2.28a, b, c, d, e, f and g) can be calculated at any given temperature.

The basic difference between the chemical reaction in Eq. (2.27) and those listed in Eqs. (2.28a, b, c, d, e, f and g) is that the former will not take place in a single step, even though it is shown as such. However, each of the seven reactions shown in Eqs. (2.28a, b, c, d, e, f and g) does occur in a single step. These are called the *elementary reactions*. Several such elementary reactions constitute a detailed kinetics mechanism for the global reaction. For elementary reactions, the molecularity is given as number of molecules participating in the reaction. The forward reactions in Eqs. (2.28a, b and d) have molecularity two, and those in other equations have molecularity one. It is not meaningful to define the molecularity for a global reaction. Also, it should be noted that the elementary reactions are reversible in general and that at a given temperature and pressure they have a given equilibrium composition, which includes reactants as well as products as listed in the reaction.

2.4.2   Reaction Mechanisms

A global reaction proceeds through several elementary reaction steps. These reactions are also known as chain reactions. These are classified as chain initiation, chain propagation, chain branching and chain termination reactions. Few of the reactions involved in the chain reaction mechanism of the hydrogen–oxygen reaction ($H_2 + 0.5\,O_2 \rightarrow H_2O$) has been used to illustrate these reactions next.

Chain initiation:

$$H_2 + M \rightarrow H + H + M$$
$$O_2 + M \rightarrow O + O + M$$

Chain propagation:

$$H + O_2 + M \rightarrow HO_2 + M$$
$$HO_2 + H_2 \rightarrow H_2O + OH$$
$$OH + H_2 \rightarrow H_2O + H$$

Chain branching:

$$H + O_2 \rightarrow OH + O$$
$$H_2 + O \rightarrow OH + H$$

Chain termination:

$$H + OH + M \rightarrow H_2O + M$$
$$H + H + M \rightarrow H_2 + M$$
$$O + O + M \rightarrow O_2 + M$$

M in the above reaction mechanism represents a collision partner, and can be any single or combination of species or even the combustor wall. Species such as H, O and OH are called radicals while HO_2 is a meta-stable species. Radicals are initially formed through chain initiation reactions. Radicals have unpaired valence electrons, which make them very reactive. Due to this, the radicals have a shorter life, when compared to other species. They react with other radicals to form covalent bonds. Chain propagation reactions produce radicals along with meta-stable and stable species. On the other hand, chain branching reactions lead to a rapid production of radicals alone, which causes the overall reaction to proceed almost explosively fast. The reaction comes to a completion through chain termination reactions, in which the radicals recombine to form the final products.

In premixed flames, where chemical kinetics plays an important role, all the reaction pathways should be understood and a detailed mechanism should be employed for analysis. For instance, even for a well-known reaction such as the hydrogen–oxygen reaction, a reaction mechanism having at least 20 steps is required to completely analyze premixed flame propagation. More elaborate reaction mechanisms with hundreds of steps are also available, and these are essential for predicting highly transient phenomena such as ignition and extinction. For instance, for the simplest hydrocarbon, methane, a reaction mechanism with more than 350 reaction steps is available. It should be noted that a heat of the reaction calculated taking into account of reactions in a detailed reaction mechanism will be same as that calculated using a global single-step reaction according to the Hess law of summation.

2.4.3 Rate Equations

Consider a general global reaction given by $F + a\text{Ox} \rightarrow b\text{Pr}$, where F is the fuel, Ox is the oxidizer and Pr represents the products. The rate (in kmol/m^3 s) at which the fuel is consumed can be expressed as,

$$\frac{d[F]}{dt} = -k_G(T)[F]^n[\text{Ox}]^m, \tag{2.29}$$

based on measurements in controlled experiments in the laboratory. The notation $[X]$ represents the molar concentration of species X in kmol/m^3. Equation (2.29) states that the rate of consumption of fuel (negative sign in RHS) is proportional to the concentrations of the reactants raised to some power. This is called law of mass action. The sum of the exponents, n and m, represents the overall order of the reaction. In general for hydrocarbons, the order of the global reaction varies from 1.7 to 2.2. The constant of proportionality, k_G, is called the global rate coefficient. It is a strong function of temperature and can be determined only from experiments.

For an elementary reaction, the law of mass action is used to arrive at the reaction rate. The notable point is that the exponents of reactant concentrations will be same as the corresponding stoichiometric coefficients. For example, for a bimolecular reaction represented by $A + B \rightarrow C + D$, the rate in kmol/m^3 s, at which the reaction proceeds is proportional to the concentrations of A and B, given as,

$$\frac{d[A]}{dt} = -k_b(T)[A][B], \tag{2.30}$$

where k_b is the bimolecular rate coefficient. It may be noted that the order for this reaction is two, same as the molecularity. The units for k_b are m^3/kmol-s. This elementary rate coefficient is also a strong function of temperature. Although it is possible to give the elementary rate coefficients a theoretical basis based on molecular collision theory, the theory is not complete, and hence, the rate coefficient is estimated mainly based on a curve-fit of the experimental data. Most reactions encountered in the context of combustion are of second order. It is worth noting that the order of the reaction of general hydrocarbon fuels is also close to two (1.7–2.2). Dissociation reactions are unimolecular and first order in nature. Termolecular reactions with order 3 are also reported in literature, although these are relatively rare.

For both global and elementary reactions, the rate coefficients, which are strong functions of temperature, are evaluated using the curve fit of the temporal variation of concentrations of reactant and products species measured from experiments under isothermal conditions. The rate coefficient, k, is thus obtained as a function of temperature. Arrhenius suggested employing a two-parameter expression for the rate coefficient, given as,

$$k(T) = A \exp\left[\frac{-E_a}{R_u T}\right], \tag{2.31}$$

where A is referred to as the pre-exponential factor or the frequency factor. From collision theory, it is found that A varies as $T^{1/2}$. Arrhenius plots of log k versus $1/T$, using the experimental data, are employed to evaluate E_a, the activation energy. In several instances, a three-parameter functional form is extensively used to evaluate the rate coefficient, as given below.

$$k(T) = AT^m \exp\left[\frac{-E_a}{R_u T}\right]$$

The typical rate coefficients for a few bimolecular reactions in the H$_2$–O$_2$ reaction mechanism are given in Table 2.3. Exponent n used in the units of the pre-exponential factor, A, in Table 2.3, represents the reaction order.

Table 2.3 Typical values of pre-exponential factors, activation energy and temperature exponent for few reactions

Reaction	$A\left[\left(cm^3/gmol\right)^{n-1}/s\right]$	m	E_a (kJ/gmol)
$H + O_2 \rightarrow OH + O$	1.2×10^{17}	-0.91	69.1
$OH + O \rightarrow H + O_2$	1.8×10^{13}	0	0
$O + H_2 \rightarrow OH + H$	1.5×10^7	2.0	31.6
$OH + H_2 \rightarrow H_2O + H$	1.5×10^8	1.6	13.8
$H_2O + H \rightarrow OH + H_2$	4.6×10^8	1.6	77.7
$H_2O + O \rightarrow OH + OH$	1.5×10^{10}	1.14	72.2

2.4.3.1 Reversible Reactions

For a reversible reaction, the rate coefficient of the forward reaction, k_f, and that of the reverse reaction, k_r, may be related to the equilibrium constant based on the concentration, K_C. Consider an elementary bimolecular reaction

$$aA + bB \overset{k_f, k_r}{\longleftrightarrow} cC + dD.$$

At equilibrium, the rate of forward reaction is equal to that of the reverse reaction, since the product composition does not change with time. Therefore, the forward and the reverse reaction rates can be related using the law of mass action as,

$$k_f[A]^a[B]^b = k_r[C]^c[D]^d.$$

The ratio of the forward rate constant to the backward rate constant is denoted by K_C, the equilibrium constant based on the concentrations, and is given by,

$$\frac{k_f}{k_r} = \frac{[C]^c[D]^d}{[A]^a[B]^b} = K_C$$

The above equation is also referred as Guldberg and Waage's law of chemical equilibrium. The relationship between the equilibrium constants defined on the basis of species concentrations, K_C, and defined on the basis of partial pressures, K_p, is established as follows:

$$[A] = \frac{a}{V} = \frac{p_A}{R_u T} \Rightarrow K_C = \frac{\left(\frac{p_C}{R_u T}\right)^c \left(\frac{p_D}{R_u T}\right)^d}{\left(\frac{p_A}{R_u T}\right)^a \left(\frac{p_B}{R_u T}\right)^b} = K_p \left(\frac{p^0}{R_u T}\right)^{\Delta n_R}$$

In the above equation, $\Delta n_R = (c + d) - (a + b)$ and p^0 is the ambient pressure.

2.4.3.2 Net Rate of Individual Species in Multi-step Reactions

If the rates of the elementary reactions are known, the net rates of production or destruction of any species participating in a series of reaction steps could be determined. For example, considering a few forward and reverse reactions, in the H_2-O_2 system, the net rate for the participating species could be estimated as follows:

$$H_2 + O_2 \overset{k_{f1},\ k_{r1}}{\longleftrightarrow} HO_2 + H \qquad (R-1)$$

$$H + O_2 \overset{k_{f2},\ k_{r2}}{\longleftrightarrow} OH + O \qquad (R-2)$$

$$OH + H_2 \overset{k_{f3},\ k_{r3}}{\longleftrightarrow} H_2O + H \qquad (R-3)$$

$$H + O_2 + M \overset{k_{f4},\ k_{r4}}{\longleftrightarrow} HO_2 + M \quad (R-4)$$

In the elementary steps, k_{fi} and k_{ri} represent forward and reverse rate coefficients of the elementary reaction i. For example, the net rate of production of O_2 is given by the sum of the rates of all the individual steps producing O_2 minus the sum of the rates of all the individual steps consuming O_2. That is,

$$\frac{d[O_2]}{dt} = k_{r1}[HO_2][H] + k_{r2}[OH][O] + k_{r4}[HO_2][M]$$
$$- k_{f1}[H_2][O_2] - k_{f2}[H][O_2] - k_{f4}[H][O_2][M]$$

Similarly for H atom,

$$\frac{d[H]}{dt} = k_{f1}[H_2][O_2] + k_{f3}[OH][H_2] + k_{r2}[OH][O] + k_{r4}[HO_2][M]$$
$$- k_{r1}[HO_2][H] - k_{f2}[H][O_2] - k_{r3}[H_2O][H] - k_{f4}[H][O_2][M]$$

In general, a set of M multi-step reactions involving N species may be written in the following compact notation:

$$\sum_{j=1}^{N} v'_{ji} X_j \Leftrightarrow \sum_{j=1}^{N} v''_{ji} X_j \text{ for } i = 1, 2, \ldots, M$$

In the above equation, X is the concentration, v'_{ji} and v''_{ji} are the stoichiometric coefficients on the reactants and products sides of the rate equation, respectively, for the jth species of the ith reaction.

2.4.3.3 Steady-State Approximation

Radicals are rapidly reacting chemical species. Initially, when the chemical reaction starts, these radicals are formed in huge quantities through chain initiation and chain propagation reactions. Many radicals, after the initial buildup of their concentration, are consumed through the chain termination reactions at a very fast rate due to their higher concentrations. They are simultaneously produced through the chain branching reactions; however, the reactions that produce the radical are slightly slower than the reactions that consume it. As a result, at one stage during the progress of the reaction, the rate of formation of these radicals will be almost the same as their consumption rate. At this stage, the radical may be assumed to be in a steady state with a smaller concentration when compared to the concentrations of the reactants and product species. This continues until the termination of the reaction, when all the radicals are consumed and the final products are formed. By invoking this steady-state approximation for a radical, the conservation equation for the transport of the radical need not be solved even though the radical is included in the reaction mechanism. The concentration of such a radical may be calculated using the concentrations of other major species.

For example, consider the Zeldovich mechanism for NO formation given by,

$$O + N_2 \xrightarrow{k_1} NO + N$$

$$N + O_2 \xrightarrow{k_2} NO + O$$

The first reaction in which N is produced is slower than the second one in which N is consumed. The net production of N atoms is given by,

$$\frac{d[N]}{dt} = k_1[O][N_2] - k_2[N][O_2].$$

After an initial transient period, where N is produced through the first reaction, it is consumed by the second reaction at a rate faster than the rate it is produced. This is because of the fact that the rate of the second reaction depends upon the concentration of N. As a result, the concentration of N in the mixture decreases and so the reaction rates of first and second reactions tend to become the same. Therefore, the net rate of production of N becomes zero; i.e., $d[N]/dt$ approaches zero. The steady-state concentration of $[N]_{ss}$, where subscript ss denotes steady state, is given by,

$$0 = k_1[O][N_2] - k_2[N]_{ss}[O_2],$$
$$\Rightarrow [N]_{ss} = \frac{k_1[O][N_2]}{k_2[O_2]}.$$

2.4.3.4 Partial Equilibrium Treatment

In a chemical reaction mechanism, there are fast as well as slow reactions. Chain propagating and chain branching reactions are fast reactions in which usually both forward and reverse reactions occur at high rates. For example, in the reaction $O_2 \leftrightarrow 2O$, the forward and reverse reactions are extremely rapid, such that this reaction attains equilibrium well before other reactions go toward their completion. That is, the slow reactions, which can be termolecular combination reactions, are the rate determining or the rate limiting reactions. Therefore, some of the fast reactions proceed so fast in comparison that they may be considered to be in equilibrium. This approach is known as partial-equilibrium treatment. As in the case of steady-state approximation, this approach also eliminates the need for solving the conservation equations for many radicals and the concentrations of these radicals are written in terms of stable and/or meta-stable species.

Consider the following reactions,

$$H_2 + O_2 \xrightleftharpoons[k_{r1}]{k_{f1},\ k_{r1}} 2OH,$$

$$H_2 + OH \xrightarrow{k_2} H_2O + H$$

The production of radical OH in the first reaction and its recombination to form the reactants is considered to be in partial equilibrium. Therefore,

$$K_p = \frac{p_{OH}^2}{p_{H_2}p_{O_2}} = \frac{[OH]^2}{[H_2][O_2]} \Rightarrow [OH] = \left(K_p[H_2][O_2]\right)^{0.5}$$

The rate of production of H_2O is given by,

$$\frac{d[H_2O]}{dt} = k_2[OH][H_2] = k_2 K_p^{0.5}[H_2]^{1.5}[O_2]^{0.5}$$

It is clear that the concentration of OH can be estimated using the concentrations of hydrogen and oxygen and the equilibrium constant for the reaction. Furthermore, it should be noted that, in essence, the net rate of production of H_2O is obtained using the concentrations of the reactants, H_2 and O_2, and a global single-step rate parameter, namely, $k_G = k_2(K_p)^{0.5}$, has been obtained. The steady-state approximation and partial equilibrium approach together may be applied to reduce the conservation equations for certain species involved in a detailed chemical kinetics mechanism. The added advantage, apart from a reduction of the number of governing equations, is that short, reduced or even single-step mechanisms may be proposed.

Review Questions

1. How is equivalence ratio calculated?
2. What is absolute enthalpy?

3. Find the difference between the adiabatic flame temperature for methane–air reaction occurring in constant pressure and constant volume chambers.
4. Write the condition for chemical equilibrium for a constant volume combustion process.
5. What is Le Chatelier principle?
6. Write the differences between a global reaction and an elementary reaction.
7. What are the types of chain reactions?
8. What is law of mass action? How it is applied to hydrogen–air reaction?
9. Write an expression for rate coefficient following Arrhenius method.
10. Write the relationship between equilibrium constant based on partial pressures and concentrations.
11. Express a multi-step reaction in a compact notation.

Exercise Problems

1. Calculate the mass of theoretical air required to burn 1 kg of biogas (60% CH_4, 30% CO_2 and 10% N_2).
2. Exhaust gas composition from an n-heptane fueled combustor is measured as follows: oxygen occupies 5.5% by volume, volumetric ratio of CO to CO_2 is 0.17, and remaining is nitrogen. Determine the equivalence ratio of the reactant mixture. Also, evaluate the molar concentration of the products and the total product volume resultant from burning 1 kmol of the fuel. Assume the products to be at atmospheric pressure and at an average temperature of 1350 K.
3. Calculate the standard heat of reaction for an equimolar mixture of CO and H_2 burning with 120% theoretical air in a constant volume combustion chamber.
4. Ethanol (C_2H_5OH) is burnt in a furnace with equivalence ratio of 1.1. If no hydrogen or oxygen is found in the exhaust, determine the product composition per kmol-fuel. For the constant pressure process, determine the standard heat of combustion per kg of fuel. Take standard heat of formation of ethanol as $-$ 235,310 kJ/kmol.
5. A liquid alkane, C_nH_{2n+2}, is burned with dry air and the product composition is measured on a dry mole basis as: 9.6% CO_2, 7.3% O_2 and 83.1% N_2. Find the standard heat of reaction per kg of fuel and per kg of oxygen. Also, find the higher heating value.
6. Carbon is burned with air in a furnace with 150% theoretical air at a constant pressure of 1 atm and both reactants are supplied at 298 K. What is the adiabatic flame temperature?
7. Consider the reaction: $CO_2 \leftrightarrow CO + \frac{1}{2}O_2$. At 100 atm, the mole fraction of CO is 0.0289. Determine K_p.
8. Consider the combustion products of decane ($C_{10}H_{22}$) with air at an equivalence ratio of 1.25, pressure of 1 atm, and temperature of 2200 K. Estimate the mixture composition assuming H_2O, CO, CO_2 and H_2 in the products, apart from nitrogen.
9. Consider Zeldovich mechanism for NO formation given by,

$$O + N_2 \xrightarrow{k_1} NO + N$$

$$N + O_2 \xrightarrow{k_2} NO + O$$

Assuming N in steady state and $O_2 - O$ in partial equilibrium, express the concentrations of N and O as a function of other species, rate coefficients and equilibrium constant. Further, express the net formation rate of NO as a global step involving N_2, O_2, k_1, k_2 and equilibrium constant.

10. Reverse reaction rate of an elementary reaction, $NO + O \leftrightarrow N + O_2$, at 2300 K is 5.28×10^{12} cm^3/gmol-s. Using the thermodynamic approach determine its forward reaction rate.

Chapter 3
Review of Combustion Phenomena

In the first chapter, different modes of combustion were briefly discussed. The volumetric combustion process, explosions and other non-flame modes of combustion are encountered only in specific applications. In general, flame mode of combustion occurs in burners used in domestic and industrial applications, irrespective of whether they are designed to utilize gas, liquid or solid fuels. It is thus important to understand the physics of different types of flames that are likely to occur in such burners. In this chapter, homogeneous flames, which involve gaseous fuel and oxidizer, and heterogeneous flames, which involve condensed fuels such as liquid and solid fuels, are described. Their characteristics, physical processes and controlling factors, such as temperature, pressure, equivalence ratio and oxygen concentration, are presented in detail.

3.1 Premixed Flames

In an isolated system, when the fuel and oxidizer are mixed in proper proportions and ignited using a localized high-temperature source, *sustained reaction* may take place. Such a reactant mixture, in which a flame can be initiated, is called a *flammable mixture*. In other words, any reactant mixture, which can burn without additional oxidizer or fuel, is called a flammable mixture. If the amount of fuel or oxidizer present in the mixture is too much or too little, then sustained reaction may not happen in that mixture in spite of ignition. A reactant mixture, based on the type of fuel or oxidizer used, will be flammable only in certain proportions and under certain conditions of temperature and pressure. The heat released during combustion will also vary based on the composition of the mixture. Limits of flammability are determined by using standard test procedures. Table 3.1 presents the *lower* (lean mixture that can ignite) and *upper* (rich mixture that can ignite) *flammability limits* for some fuels calculated based on the ability of a flame to propagate in the mixture after

V. Raghavan, *Combustion Technology*,
https://doi.org/10.1007/978-3-030-74621-6_3

Table 3.1 Flammability
limits of some selected fuels

Fuel	Volumetric % of fuel		% fuel in stoichiometric fuel
	Lower	Upper	
Methane	5.0	15.0	9.47
Propane	2.37	9.5	4.02
Butane	1.86	8.41	3.12
Heptane	1.0	6.0	1.87
Ethylene	2.75	28.6	6.52
Acetylene	2.5	80.0	7.72
Benzene	1.4	6.75	2.7
Hydrogen	4.0	74.2	29.5
Carbon monoxide	12.5	74.2	29.5
Methanol	6.72	36.5	12.2
Ethanol	3.28	18.95	6.52

ignition. These are estimated using an experimental setup consisting of an upward oriented 50 mm glass tube filled with reactant mixture at atmospheric pressure. The percentage of fuel in the stoichiometric mixture is also reported in Table 3.1. It may be noted that fuels such as hydrogen, acetylene and to some extent CO have a wide range of flammability limits.

When a flammable mixture is available in a combustion chamber, a flame may be formed and sustained inside the combustion chamber. Such flames are called *premixed* flames. These flames may either be propagating into the reactant mixture or be stationary when the reactant mixture is supplied to the flame zone at a certain speed. This speed is one of the important characteristics of premixed flames and is called the *laminar flame speed*.

3.1.1 Laminar Premixed Flames and Flame Speed

Consider a long duct open at either one or both of its ends, filled with a flammable reactant mixture at a certain temperature. When this mixture is ignited at an open end, a premixed flame will be initiated. Initial *piloted ignition* will depend upon the *magnitude of the energy source* used as well as the *volume of the reactant mixture* available. A minimum amount of ignition energy and a critical volume of reactant mixture are required for ignition. These quantities depend on the initial temperature and composition of the reactant mixture. Once ignited at its open end, a flame will steadily propagate at a subsonic speed. The hot products of combustion leave the duct through the open end, where the mixture has been ignited. This process is called

Fig. 3.1 Propagating
flame—illustration of a
deflagration

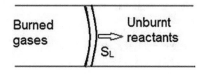

deflagration, and the propagation speed is called the laminar flame speed, S_L. This is illustrated in Fig. 3.1.

Laminar flame speed is the speed at which the flame steadily consumes the reactant mixture. When the reference frame is attached to the flame itself, the speed with which the reactant mixture approaches the flame in the direction normal to it is the laminar flame speed. While it is possible to measure this speed, it is not a fundamental property of the fuel. It may be noted that due to wall effects the flame shown in Fig. 3.1 is curved. The laminar flame speed depends on several factors such as reactant equivalence ratio, its temperature and pressure. Further, heat and radical losses to the chamber walls affect the rate of flame propagation. Furthermore, the dimensions of the duct also affect the flame propagation. When the dimension of the duct is below a certain critical value, the heat release due to combustion within the small volume of the duct cannot compensate for the heat loss through the walls of the duct, and the flame will not be able to propagate through the reactant mixture. It can be noted that the temperature and the composition of the reactant species are uniform and the flame propagation depends only on the rate of reaction or the chemical kinetics. The other important characteristic of a premixed flame is the *flame temperature*. Analysis of the premixed flames requires a reaction mechanism with adequate number of species including several radicals and metastable species.

In contrast to the situation described above, if one end of the duct is closed and the mixture is ignited at the closed end, the hot product gases cannot escape out of the duct. As a result, the pressure builds up in the region between the flame and the closed end. This accelerates the flame, and under certain conditions, the flame propagation speed may even reach supersonic values. Such a high-speed flame propagation is called *detonation*. Detonation is not easy to establish in practical situations and finds place in certain unique applications as in pulse detonation engines.

In combustion chambers that employ premixed gas burners, the flame is kept stationary by supplying the reactant mixture at a certain rate, such that the magnitude of its velocity component normal to the flame surface is equal to the laminar flame speed for that mixture under that condition. A premixed flame sustained on the top of a *Bunsen burner* is illustrated in Fig. 3.2.

Typically, a *conical* premixed *luminous* flame is formed at the exit of a Bunsen burner. If U is the average velocity of the reactant mixture in the burner tube and α is the half cone angle, then the velocity component normal to the flame surface is $U \sin \alpha$. This component is equal to the average laminar flame speed, S_L. The flame speed can be determined by dividing the volumetric flow rate of the reactant mixture by the surface area of the flame. If the mixture is fuel rich, another *non-luminous non-premixed* or *diffusion* flame surrounds the conical flame. In general, in these

Fig. 3.2 Stationary laminar premixed flame from a Bunsen burner; laminar flame speed is the normal component of unburnt gas velocity. Angle α is the half cone angle of the premixed flame

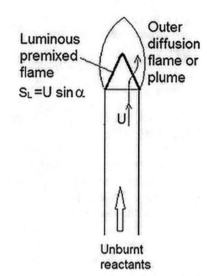

combustors, both chemical kinetics and transport processes such as convection and diffusion become important for the analysis. Photographs of methane–air premixed flames sustained at the exit of a Bunsen burner having different equivalence ratios are presented in Fig. 3.3. The temperature of the reactant mixture is kept at 297 K, and the operating pressure of the burner is around 1 bar. When the equivalence ratio is increased from 0.7 to around 1.05, the luminosity of the conical inner flame increases. This is due to an increase in the adiabatic flame temperature in these cases

Fig. 3.3 Photographs of methane–air premixed flames from Bunsen burner having different equivalence ratios

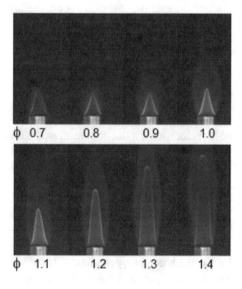

Fig. 3.4 Variation of adiabatic flame temperature as a function of equivalence ratio for methane–air mixture, calculated considering equilibrium products

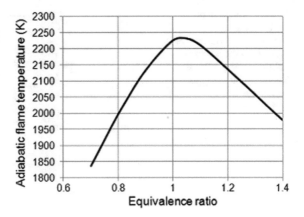

as illustrated in Fig. 3.4. Further increase in equivalence ratio results in decrease in the flame temperature. As a result, the luminosity of the inner conical flame decreases.

It is known that the standard heat of combustion, calculated at standard pressure (1 *atmosphere*) and temperature (298 K), is a constant for lean to stoichiometric mixtures. In practical constant pressure combustion chambers employing fuel lean to stoichiometric mixtures, the heat of combustion, calculated when the products exit at a temperature higher than 298 K, increases with increasing equivalence ratio and attains a maximum value for the stoichiometric mixture. However for methane–air flames, it can be seen from Fig. 3.4 that the flame temperature, which depends on the heat of reaction, attains a maximum value for a mixture having equivalence ratio slightly greater than unity. Therefore, there is a shift in the value of equivalence ratio corresponding to maximum heat of combustion and maximum flame temperature. This is due to the dependency of the thermal conductivity, specific heat and diffusivity of the gases on the temperature. Consequently, the laminar flame speed, which strongly depends on the flame temperature, also depicts a similar variation with equivalence ratio. On the richer side, both heat release and flame temperature decrease as the equivalence ratio is increased. This is shown in Fig. 3.5.

Measured values of the flame speed reported by researchers vary over a wide range. This is due to differences in the experimental setups and operating conditions they have employed. The variation trend, however, is almost the same, and the flame speed attains a maximum value when the equivalence ratio is in the range of 1.05–1.1 for methane–air mixtures.

Similar trends are observed in commonly used fuels such as acetylene, ethylene, CO, hydrogen and propane. However, the equivalence ratio at which the maximum laminar flame speed is obtained is different between these fuels. The laminar flame speeds for several fuels are comparatively shown as a function of percentage fuel in the reactant mixture in Fig. 3.6. These data correspond to Bunsen burner flames (as shown in Figs. 3.2 and 3.3). As the volumetric percentage of fuel in the reactant mixture increases, the laminar flame speed also increases and reaches a maximum

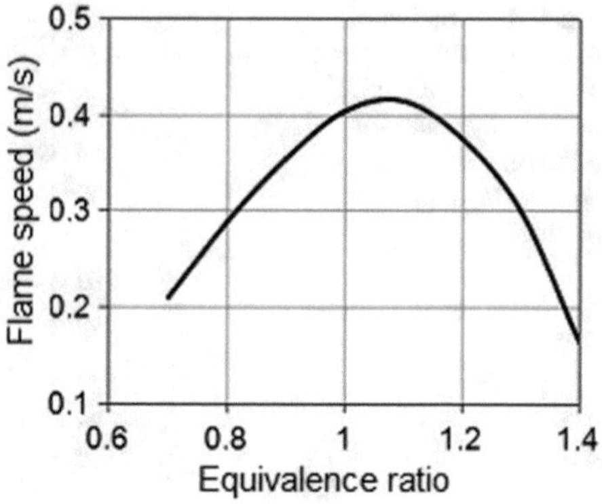

Fig. 3.5 Variation of
laminar flame speed as a
function of equivalence ratio
for methane–air mixture
estimated using a Bunsen
burner; initial reactants
supplied at 298 K,
1.01325 bar

Fig. 3.6 Variation of
laminar flame speed as a
function of volumetric
percentage of fuel in the
reactant mixture; Bunsen
burner experimental results
conducted at atmospheric
pressure

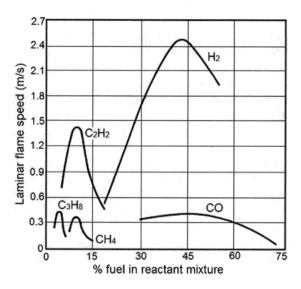

value for a particular rich mixture. Further increase in the fuel percentage causes the
laminar flame sped to decrease.

For CH_4 and C_3H_8, the equivalence ratio at which S_L becomes a maximum is
around 1.08, and for C_2H_2, it is around 1.2. On the other hand, for H_2, it is around
1.8, and for CO, it is around 2, implying that a richer mixture is required for attaining
the maximum flame speed for these fuels. One important point to note here is that the
occurrence of the maximum adiabatic flame temperature for hydrogen–air mixture
happens around an equivalence ratio of 1.07, which is almost the same as that of
methane. However, the maximum flame speed occurs around an equivalence ratio of
1.8. This is due to the variation of thermal and mass diffusivities, or their ratio called
Lewis number, *Le*, with the composition of the reactant mixture. It is also clear that
the maximum flame speed is the highest for H_2–*air* mixture, followed by C_2H_2, CO
and the straight chain saturated hydrocarbons.

3.1.1.1 Structure of a Premixed Flame

Several physical and chemical processes need to be understood in the analysis of premixed flames. Prediction of flame temperature is possible using a thermodynamic analysis and chemical equilibrium concepts, as discussed in Chap. 2. Various flame theories have been reported based on physical and chemical properties of the reactants. However, a closed-form solution for laminar flame propagation has not been possible.

Figure 3.7 presents the typical structure of a premixed flame. As the premixed reactant stream approaches a flame in the direction normal to it, due to the heat transfer from the flame, the temperature increases from the initial value of T_0 to the ignition temperature, T_i. The zone where this occurs is termed as the *preheat* zone. Adjacent to the preheat zone, the *reaction* zone exists. In general, the reaction zone has a very small thickness, of the order of a millimeter. Usually toward the end of the reaction zone, a *bright* or *luminous* zone is present, where the temperature attains a maximum, called the flame temperature, T_f. In addition to heat transfer from the reaction zone to the preheat zone, radicals such as H, O, OH and so on are also transported from the flame to the preheat zone. Reactants heated to the ignition temperature react with these radicals through the chain initiation reactions and enable the onset of chain propagation and chain branching reactions in the reaction zone. For lean to stoichiometric mixtures, this causes a reduction in the mass fraction of the reactants, $Y_{reactant}$, and a further rise in the temperature as shown in Fig. 3.7. When the reaction proceeds and the reactants are consumed, the heat release rate, $\dot{q}$, rapidly increases and attains a maximum value as shown in Fig. 3.7.

Since the major reactant species are consumed at this point, the heat release rate decreases to zero rapidly. Recombination of the species occurs downstream of the reaction zone. Depending on the initial composition of the reactant mixture, a plume of burned gas or a diffusion flame will be present in this zone. Intermediate species are formed toward the end of preheat zone and are consumed before the end of the

Fig. 3.7 Typical structure of the premixed flame

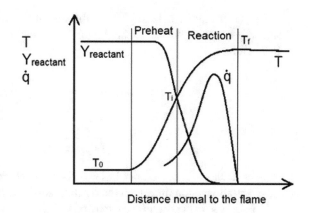

Distance normal to the flame

reaction zone. In the case of rich mixtures, the fuel that is transported out of the reaction zone burns in a diffusion flame mode.

The following theories are available for analyzing laminar flame propagation: (1) thermal theory, (2) diffusion theory and (3) comprehensive theory. In thermal theory, it is assumed that the mixture is heated by conduction to a temperature at which the rate of reaction is sufficiently rapid and self-sustaining. Heat transfer from the flame alone is assumed to be sufficient to explain the flame propagation. On the other hand, in diffusion theory, diffusion of active species such as atoms and radicals, from the reaction zone or the burned gas, into the un-reacted mixture, is assumed to initiate the reaction as well as sustain it. In reality, both diffusion of heat and diffusion of active radicals into the preheat zone contribute to flame propagation. Such a detailed analysis is employed in the comprehensive theory.

Mallard and Le Chatelier postulated a two-zone model. In zone 1, the reactant mixture is preheated to a certain temperature, called the ignition temperature, and in zone 2, all the reactions are completed. They stated that the extent of zone 1 depended on the heat conduction from zone 2, such that the interface between these zones reaches the ignition temperature. Zone 1 is similar to the preheat zone shown in Fig. 3.7. However, the slope of the temperature curve in this zone is assumed to be linear. Taking the thickness of the zone 2 (reaction zone in Fig. 3.7) to be δ, energy balance between the two zones is given by,

$$\dot{m} c_p (T_i - T_0) = \lambda \frac{(T_f - T_i)}{\delta} A, \qquad (3.1)$$

where λ is the thermal conductivity, $\dot{m}$ is the mass flow rate of the mixture, T_0 is the temperature of the unburned gases and A is the cross-sectional area, taken as unity. If the flame propagation is assumed to be one-dimensional, the mass flow rate may be evaluated as,

$$\dot{m} = \rho U A = \rho S_L A.$$

Here, ρ is the average density and U the velocity of the reactant gases. Since the unburned gases enter normal to the flame (streamline in Fig. 3.2), by definition, $S_L = U$. Therefore,

$$\rho S_L c_p (T_i - T_o) = \lambda \frac{(T_f - T_i)}{\delta},$$

$$S_L = \frac{\lambda}{\rho c_p} \frac{(T_f - T_i)}{\delta (T_i - T_o)}$$

If ω represents the reaction rate based on fractional conversion of the reactant and τ is the overall reaction time, then the reaction zone thickness, δ, is given by $\delta = S_L \times \tau \approx S_L/\omega$. The expression for S_L may be written as,

$$S_L \approx \frac{\lambda}{\rho c_p} \frac{\omega(T_f - T_i)}{S_L(T_i - T_o)},$$

$$S_L \approx \left(\frac{\lambda}{\rho c_p} \frac{(T_f - T_i)}{(T_i - T_o)}\omega\right)^{1/2} \approx (\alpha\omega)^{1/2} \tag{3.2}$$

The laminar flame speed thus depends upon the square root of the product of thermal diffusivity and the reaction rate. Due to the exponential dependence of the reaction rate on temperature, the temperature dependence of the laminar flame speed may be expressed as,

$$S_L \approx \left[\exp\left(-\frac{E_a}{R_u T}\right)\right]^{1/2}$$

For an nth-order reaction, the reaction rate depends on pressure, p, as p^{n-1}. This implies that the pressure dependence of the laminar flame speed may be expressed as,

$$S_L \approx \left(\frac{1}{\rho}p^{n-1}\right)^{1/2} \approx \left(p^{n-2}\right)^{1/2}.$$

This states that the flame speed is independent of pressure for second-order reactions. Most hydrocarbon–air reactions have an overall reaction order close to two. Therefore, the flame speeds for hydrocarbons are generally pressure independent. However, experiments have shown that S_L decreases slightly in a linear manner with pressure.

The reaction zone thickness may be written as $\delta = \alpha/S_L$. Most hydrocarbons have a maximum flame speed of about 0.4 m/s. With the value of thermal diffusivity evaluated at a mean temperature of about 1300 K, a value for δ close to 0.1 cm is obtained. Therefore, the reaction zone thickness is of order of a millimeter.

Investigators such as Zeldovich, Frank-Kamenetskii and Semenov included chemical reaction in their energy balance equation. Ignition temperature was assumed to be very close to the flame temperature. This assumption is logical in the case of hydrocarbon flames, where the major heat release is due to CO oxidation and this depends on the availability of the OH radical. By the time the CO and OH reaction takes place, temperature in the interface of the two zones would have reached a value almost that of the flame temperature. Interestingly, their analysis that also included the species diffusion resulted in an expression for S_L, almost identical to that in Eq. (3.2).

Tanford and Pease analyzed flame propagation using the diffusion theory. Their system consisted of several chain-carrying species. Convection and diffusive transports of all these species were considered in their analysis. They showed that the laminar flame speed depended on the square root of the product of the reaction rate and mass diffusion coefficient, instead of the thermal diffusivity proposed by the thermal theory.

A simplified mathematical model has been proposed by Spalding to estimate the laminar flame speed. The one-dimensional model assumed a Lewis number of unity, constant specific heats and a global one-step reaction given by,

$$1 \text{ kg fuel} + s \text{ kg oxidizer} \rightarrow (1 + s)\text{kg products}. \tag{3.3}$$

With this model, the expression for laminar flame speed is given as,

$$S_L = \left(2\alpha[1 + s]\frac{(-\omega_F)}{\rho_u}\right)^{1/2}. \tag{3.4}$$

In Eq. (3.4), α is thermal diffusivity, ω_F is rate of consumption of the fuel in kg/m^3 s, and ρ_u is the density of the unburned mixture. The negative sign is used to consider only the magnitude of net fuel consumption rate. It is clear that the laminar flame speed depends on square root of the product of thermal diffusivity and the reaction rate.

In the comprehensive theory, nonlinear species conservation and energy conservation equations are solved using temperature-dependent mixture properties. Finite rate chemical kinetic mechanisms are employed. Several studies have been carried out on understanding the structure of premixed flames using detailed chemical kinetics mechanisms. Software such as Chemkin and FlameMaster has numerical models to estimate laminar premixed flame structure using detailed chemical kinetics mechanisms, variable thermo-physical properties and radiation models.

Different experimental techniques are available to measure laminar flame speed. Bunsen burner, flat flame burner, spherical combustion vessel, constant heat flux burner and flame propagation in long tube are some of the widely used methods. However, global empirical correlations for laminar flame speed, valid for several fuels under several operating conditions, have not been possible because of the dependency of the laminar flame speed on many parameters and operating conditions.

3.1.2 Basic Characteristics of Turbulent Premixed Flames

Turbulent flows are chaotic in nature. Due to random temporal oscillations in all the variables, turbulent flames are highly oscillatory. Consider a laminar premixed flame, which has a thin flame zone. When the flow field becomes turbulent, where eddies of different sizes are involved, due to the interaction between eddies and the flame zone, a simple conical premixed flame becomes convoluted. The regimes of turbulent premixed flames are based on the magnitudes of reaction rate, turbulent intensity and length scales involved in the turbulent flow. Consider a case where the reaction rates are much higher than the turbulent mixing rates. Here, the flame zone is thin and

Fig. 3.8 Superimposition of turbulent flame images and typical view of a time-averaged turbulent flame image

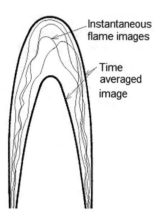

Instantaneous flame images

Time averaged image

eddies interact with this thin flame zone. From experiments, thin reaction zones can be recorded using a high-speed camera to visualize the flame. An instantaneous flame image is highly convoluted because of the turbulent oscillations. Figure 3.8 shows superimposed instantaneous flame images, as well as a time-averaged flame image, which bounds all the instantaneous images. The time-averaged image represents a thick reaction zone, where eddies contribute to mixing as well as reaction processes over a given time period. The instantaneous images show relatively thin reaction zones as in a laminar flame, and these reaction zones are generally called laminar flamelets.

Turbulent flows involve different levels of length scales. First one is the characteristic length involved in the flow (L). Second scale is the integral scale, called the turbulent macroscale (l_0). This represents the mean size of the medium- to large-sized eddies in the turbulent flow and is calculated by integrating the correlation coefficient of velocity fluctuations measured at two locations at the same time instant. Third length scale is called the Taylor microscale (l_λ). This is an intermediate scale weighted more toward smaller eddies and is the ratio of the root mean square value of velocity fluctuation $\left(v'_{rms}\right)$ to the mean strain rate. Finally, the Kolmogorov scale (l_K) is the one that represents the size of the smallest eddy in the flow. At this scale, viscous effects are predominant and it is calculated as a function of kinematic viscosity and rate of dissipation of turbulent kinetic energy. Based on these length scales and the turbulent intensity, which is based on the value of v'_{rms}, turbulence Reynolds numbers may be defined as follows:

$$\text{Re}_{l_0} = v'_{rms} l_0 / v, \ \text{Re}_{l_\lambda} = v'_{rms} l_\lambda / v, \ \text{Re}_{l_K} = v'_{rms} l_K / v. \tag{3.5}$$

Here, v is the kinematic viscosity. The ratio of the length scales may be estimated as a function of turbulence Reynolds numbers as follows:

$$l_0 / l_K = (\text{Re}_{l0})^{0.75} \text{ and } l_0 / l_\lambda = (\text{Re}_{l0})^{0.5}.$$

Regimes of premixed turbulent flames may be understood by comparing the laminar flame thickness, δ, with the different turbulent length scales. Three regimes of turbulent premixed flames have been observed experimentally by various researchers, and these are listed as follows:

Wrinkled laminar flames: $\delta \leq l_K$.
Flamelets in eddies: $l_0 > \delta > l_K$.
Distributed reactions: $\delta > l_0$

One more important dimensionless number, *Damköhler number*, Da, is defined as the ratio of characteristic flow time to characteristic chemical time. Chemical time can be estimated from laminar flame speed as δ/S_L. Since turbulent flow involves eddies of several sizes, there are multiple timescales associated with these eddies. Using turbulent macroscale, the flow time may be estimated as l_0/v'_{rms}. Thus, Da can be expressed as,

$$\mathrm{Da} = \frac{l_0/v'_{rms}}{\delta/S_L} = \left(\frac{l_0}{\delta}\right)\left(\frac{S_L}{v'_{rms}}\right) \tag{3.6}$$

When the reaction rates are faster than turbulent mixing rates, then Da will be greater than 1. If Da $\gg$ 1, then the reaction rate may be assumed to be infinitely fast. On the other hand, if the turbulent mixing rates are much higher than the reaction rates, then Da $\ll$ 1. Further, Eq. (3.6) may be viewed as the ratio of length-scale ratio (l_0/δ) to relative turbulent intensity (v'_{rms}/S_L). It may be noted that if the length scale is fixed, then Da value decreases as the turbulent intensity is increased. In turn, turbulent intensities depend upon the turbulence Reynolds numbers.

When both l_0/δ and v'_{rms}/S_L are less than 10, the flame is phenomenally *laminar*. When v'_{rms}/S_L is less than unity, for a wide range of length-scale ratio greater than unity, *wrinkled flame* regime is established. In this regime, Da $\gg$ 1 and eddies are larger than the reaction zone thickness as the reaction rates are much faster. Therefore, eddies cannot penetrate into the reaction zone and the flame surface becomes wrinkled. This is shown schematically in Fig. 3.9. Wrinkling causes an increase in the overall surface area of the flame, and as a result, the flame speed increases. In this regime, the ratio of turbulent flame speed to the laminar flame speed is proportional to the relative turbulent intensity.

When v'_{rms}/S_L is greater than 1, for almost the same range of length-scale ratio as in the above regime, *flamelets in eddies* regime are established. Damköhler numbers are greater than unity and have moderate values in this regime. Schematically, this type of flame zone is shown in Fig. 3.10. There are *pockets* of unburned gases in the reaction zone along with almost fully burned gas. The rate at which these pockets of unburned gases burn and reduce in size is determined by *turbulent mixing rates and chemical reaction plays a lesser role*. When Da approaches 1, the flame zone becomes thicker and local flame extinction and re-ignition may take place in such cases.

When the length scale is smaller and relative turbulent intensity is larger, *distributed reaction* regime is established. This zone has turbulent eddies present

Fig. 3.9 Schematic of a wrinkled flame

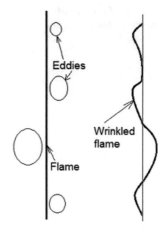

Fig. 3.10 Illustration of flamelets in eddies regime

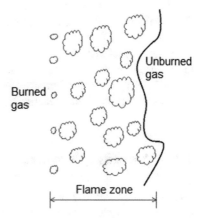

within the reaction zone as shown in Fig. 3.11. In this regime, Da < 1, and therefore, the reaction times are larger than the turbulent mixing times. This causes fluctuations in all the variables such as temperature and species concentrations and, therefore, in the reaction rates as well.

Fig. 3.11 Schematic of distributed reaction zone

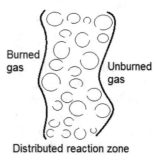

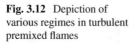

Fig. 3.12 Depiction of
various regimes in turbulent
premixed flames

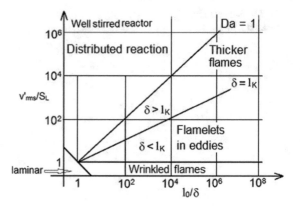

Figure 3.12 presents a summary of the turbulent premixed flame regimes in a relative turbulent intensity and length-scale plot. Similar plot in Da and turbulence Re domain may also be generated. Wrinkled flames and flamelets in eddies are formed when Da ≫ 1. Flamelets in eddies are formed when $\delta < l_K$. Thick wrinkled flame with a somewhat distributed reaction zone is formed when Da is slightly greater than unity. In this regime, $\delta > l_K$ and local flame extinction and re-ignition may occur. When Da < 1, well-stirred process takes place due to strong interaction between the eddies and the distributed reaction zone.

3.1.3 Stability of Premixed Flames

Two important stability problems arise in burners with premixed flames. Consider a Bunsen burner and a premixed flame established over the burner rim. In a steady condition, the flame anchors on the burner rim. In fact, the flame anchors a few millimeters away from the top of the rim due to the heat and radical losses from the flame to the wall. At this stable condition, the premixed reactant is continuously fed at a certain rate and the flame sits close to the burner exit and consumes the reactants. It is possible to attain this condition only within a narrow range of reactant flow rates for a given reactant composition and temperature. When the reactant flow rate is increased, it is possible for the reactant flow speed at a location to exceed the laminar flame speed. Then, the unburnt reactant can displace the flame anchoring farther away from the burner rim and this process is called *lift-off*. When the flame lifts off, it may become noisy and may also *blow off* completely. This blow-off is caused due to the dilution of the mixture by the entraining atmospheric air into the lifted flame. Further, the entrained cold air can quench the flame. This is the first instability related to premixed flames.

On the other hand, if the flow rate of the reactant (keeping the composition the same) is decreased from the value corresponding to the stable condition, the flame which is anchored above the rim of the burner will not be able to receive the reactant

Fig. 3.13 Illustration of
lift-off and flashback
phenomena

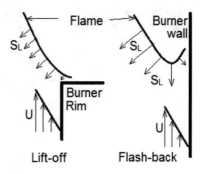

at the rate at which it is consumed. As observed in the case of deflagration in a long
tube, the flame now starts to travel upstream, into the burner, consuming the unburnt
reactants. This is possible only if the burner diameter is larger than a critical value
called *quenching distance* (discussed in the next section). This process is called a
flashback. Flashback is more dangerous than lift-off, as the flame that travels within
the burner may ignite a relatively larger volume of the reactants if present. This might
even lead to an explosion hazard.

Both these instabilities are caused due to a mismatch between local laminar flame
speed and the reactant velocity. These are illustrated in Fig. 3.13. The flame anchors
near the rim, where the laminar flame speed is decreased due to heat and radical
losses from the flame to the wall. The laminar flame speed, S_L, increases and reaches
a constant value along the flame, as illustrated in Fig. 3.13 (left). The profile of the
unburnt gas velocity, U, close to the wall is also shown. Lift-off is caused when U
exceeds the value of S_L, at several locations. The flame lifts off and sustains at a
particular position, where the local Da is close to unity. The height measured from
the burner rim to the lifted flame base is called the lift-off height. Further increase
in U causes the lift-off height to increase, and finally after reaching a critical value,
the flame blows off. When the reactant velocity is reduced below the local flame
speed, the flame propagates into the burner having a certain minimum diameter.
This is schematically shown in Fig. 3.13 (right), where the profile of S_L of a flame
propagating inside a tube and that of U are shown.

While designing premixed burners for a range of composition and temperature of
the reactant mixtures, safe operating conditions should be clearly indicated. Operating conditions such as air and fuel flow rates are customarily plotted on a stability
map in order to delineate the limits for flashback, lift-off and blow-off, and the
design area is also depicted. Figure 3.14 shows a typical stability map. Even though
the design area may be large, the actual operation is restricted to a small area keeping
issues such as transients in the flow rates in mind.

Fig. 3.14 Typical flame
stability map indicating the
limits of flashback and
lift-off, and the design area
considered for the burner
operation

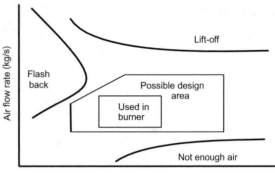

3.1.4   Quenching Distance

As mentioned earlier, when a stable flame is established on a typical Bunsen burner, and the mixture flow rate is decreased below a certain value for a given composition, the flame may flash back into the burner, provided the burner diameter is greater than a certain value. The maximum value of the burner diameter, for which the flame cannot propagate inside the burner in the flashback regime, is called *quenching distance*. That is, the flame is quenched as it tries to propagate inside the burner. When the flame propagates inside a burner, there is a certain amount of volumetric heat release due to the exothermic reaction. Also, there is heat loss from the flame to the burner wall. When the diameter of the burner is decreased, the heat released within the reduced volume also decreases. Even though the surface area of the burner has reduced, for a critical value of the burner diameter, the heat lost to the walls equals the heat released during the reaction. At this condition, the flame is quenched. Radical losses to the burner wall can also quench the flame. In general, the quenching distance is proportional to the flame thickness. Table 3.2 presents the quenching distances for several fuels when the reactant mixture is stoichiometric. It may be noted that stoichiometric hydrogen–air mixture has a small quenching distance.

Table 3.2 Typical values of
quenching distances for
stoichiometric fuel–air
mixture constituted by
different fuels

Fuel	Quenching distance (mm)
Methane	2.5
Propane	2.0
Acetylene	2.3
Ethylene	1.3
Hydrogen	0.64
Methanol	1.8
n-Decane	2.1

3.2 Non-premixed Flames

As the name suggests, these flames are formed by separately supplying the fuel and oxidizer into the combustion chamber. Here, a reaction zone is formed at locations where the fuel and oxidizer are almost mixed stoichiometric proportion. Flame zone prevails in the interface between the fuel and oxidizer, with very little leakage of reactants across the flame zone. This type of flame is also called a *diffusion flame*, because the flame location is controlled by the diffusion of the reactant species. Since the transport processes are slower than a typical combustion reaction, chemical kinetics has little role to play in diffusion flames. Based on the configurations of the fuel and oxidizer supply, non-premixed flames may be classified as jet, co-flow, counterflow and cross-flow diffusion flames. In a jet diffusion flame, fuel alone comes out of a burner port into a quiescent environment of air/oxidizer. Due to the momentum of the fuel jet, ambient air is entrained and mixes with the fuel. In a co-flow diffusion flame, an internal jet of fuel is surrounded by an external annular jet of oxidizer. Fuel and oxidizer diffuse in the radial direction and mix with each other. In an opposed or a counterflow diffusion flame, two opposing coaxial jets of fuel and oxidizer create a flat flame around the stagnation zone. In cross-flow diffusion flames, the fuel and oxidizer streams are supplied at right angles to each other, and a boundary-layer-type flame is formed. The flame locations in all these cases are controlled by the associated transport processes.

3.2.1 Laminar Jet Diffusion Flames

Typically, jet diffusion flames are formed in a typical Bunsen burner, when air entrainment within the burner tube is prevented and only fuel emerges out the burner. The fuel jet entrains air from the atmosphere and spreads along the radial direction. Oxygen from the ambient air and the fuel mixes, and when ignited, a flame zone is formed at locations where the mixture composition is around the stoichiometric value. These types of flames are characterized by their visible length, measured from the burner exit, that indicates the distance the fuel has to travel to find the required amount of the oxidizer. The fuel molecules travel through molecular diffusion in the radial direction and predominantly by convection in the axial direction. Within the laminar regime, as the fuel flow rate is increased, the flame length increases. This is demonstrated in Fig. 3.15, where instantaneous photographs of Liquefied Petroleum Gas (LPG) jet diffusion flames are shown. While premixed flames display bright blue and non-luminous blue colors as shown in Fig. 3.3, diffusion flames, on the other hand, display a range of colors including bright yellow or orange color. It is clear that the flames exhibit dull blue color, almost non-luminous, near the burner rim, where they anchor. This is where the fresh air from ambient mixes with the emerging fuel jet. Further upward, soot inception and its growth take place. The bright emission arises from soot radiation. Laminar jet diffusion flames are quite steady, even though

Fig. 3.15 Photographs of
laminar LPG jet diffusion
flames for different fuel flow
rates

2.7   3.3   3.9   5.6

Liters per hour

tip oscillations are observed above a certain fuel flow rate. In general, the operating range of a jet diffusion flame is quite high when compared to the Bunsen burner premixed flame. However, soot formation, CO and unburnt hydrocarbon emissions may be present in these types of flames. Based on the fuel and oxidizer supply, soot incepted may not be oxidized and soot particles may leave the flame tip as smoke.

At a particular fuel flow rate, a laminar jet diffusion flame transitions to a turbulent flame. Turbulent jet diffusion flames are highly oscillatory just like the turbulent premixed flames and are noisy too. An interesting behavior of turbulent jet diffusion flame is that for a given burner port diameter, and the flame height remains almost unaltered when the fuel flow rate is increased further. Characteristics of turbulent diffusion flames are discussed in a later section.

3.2.1.1   Structure of a Laminar Jet Diffusion Flame

The structure of a methane jet diffusion flame, predicted using a numerical model, is shown in Fig. 3.16. The fuel jet evolves from a burner having a circular cross section with radius, R. Ambient air is entrained as a result of transfer of momentum from the fuel jet due to viscous action, and the entrained air mixes with the fuel. At locations where fuel and air are mixed in stoichiometric proportions, upon ignition, a thin flame zone is formed (contour line of $\phi = 1$ in Fig. 3.16). The flame length, L_f, is

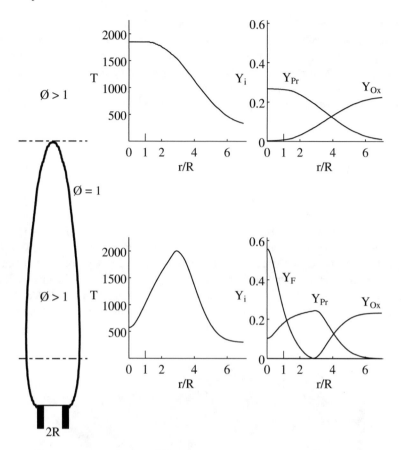

Fig. 3.16 Predicted structure of a jet diffusion flame

the axial distance measured from the burner exit to the tip of the contour line, $\phi = 1$. The radial profiles of temperature and species mass fractions describe the structure of the diffusion flame. Radial profiles of these variables drawn at two axial locations, one above the burner exit and another at the flame tip, are shown in Fig. 3.16. At the lower section, the temperature increases from a value of around 600 K at the axis and reaches a maximum value ($T_f = 1850$ K) at a radial location, r_f, called the flame radius, where $\phi = 1$. It decreases rapidly with increasing radius and asymptotically approaches the value of the ambient temperature, 300 K. The fuel (F) from the jet side diffuses toward the flame zone in the radial direction, predominantly due to the concentration gradient and also due to temperature gradient. At the flame zone, it is consumed almost completely. Similarly, the oxidizer (Ox) diffuses toward the flame zone from the ambient. It is also consumed almost completely in the flame zone. It may be noted that there is a small leak of both fuel and the oxidizer through the flame zone. On the other hand, the products (P) are formed around the flame zone and diffuse toward both the jet and the ambient sides.

At the flame tip, it is clear that the fuel does not exist. Fuel has traveled up to this point in the axial distance and it has been completely consumed at the flame tip. At this point, the required oxygen has come from the ambient as seen in the radial profile. Temperature reaches its maximum value at the flame tip, and it decreases in the radial direction and asymptotically reaches the ambient temperature value. Products, formed at the flame tip, diffuse toward the ambient.

3.2.2 Turbulent Diffusion Flames

Turbulent jet diffusion flame is obtained when the fuel flow rate is increased beyond a critical value. As mentioned earlier, tip oscillations are observed in laminar diffusion flames, once the fuel flow rate is increased beyond the critical value. As the fuel flow rate is further increased, these tip oscillations propagate upstream, creating fluctuations on the flame sheet. A smooth laminar flame surface gradually transitions to a highly oscillatory turbulent flame surface, as shown in the instantaneous flame photographs in Fig. 3.17. The oscillations are due to the interaction of turbulent eddies of different scales with the flame zone. Also, the molecular-level mixing process in a laminar flame is highly enhanced due to the turbulent eddies. Therefore, once the jet

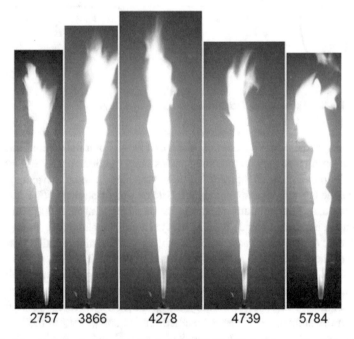

Fig. 3.17 Instantaneous photographs of turbulent LPG jet diffusion flames for different fuel jet Reynolds numbers

flow becomes fully turbulent, the diffusion flame length remains almost a constant. It may be noted that the flame lengths are almost the same for a wide range of Reynolds numbers in Fig. 3.17. Further increase in the fuel flow rate results in an increase in the noise level of the flame. Also, at another critical fuel flow rate, the flame lifts off from the burner and sustains at a certain height from the burner exit. After this point, when the fuel flow rate is further increased, the lift-off height gradually increases and the flame eventually blows off.

In fact, when a fuel jet emerges out from a very small hole, the jet velocity may be very high. Under such conditions, even when the Reynolds number, calculated based on the diameter of the port and jet velocity, is less than the general critical value, the flame can lift off in the laminar regime itself. For instance, LPG jet emerging out of a small hole of around 0.5-mm diameter can lift off at a Reynolds number of around 1900, which occurs when the fuel jet velocity is around 15 m/s. The process of flame lift-off is due to a very limited flow residence time within which the chemical reaction cannot be completed. A lifted diffusion flame is like a partially premixed flame, as a certain amount of ambient air has mixed with the fuel before the mixture reaches the flame zone. Lifted flames are noisy and are also quite unstable. It should be noted that the lift-off is the only instability observable in diffusion flames.

3.2.3 Flame Height Correlations

A diffusion flame is characterized by its visible length. A simple scaling analysis to understand the factors on which the length of a diffusion flame depends upon is presented next.

When the reaction rates are much higher than the rates of diffusion and convection processes, the resultant reaction zone has almost zero thickness. For a jet diffusion flame, molecular diffusion is predominant along the radial direction and convection occurs along the axial direction. In the laminar regime, due to molecular diffusion, if the fuel molecules travel a distance y in the radial direction, then in terms of molecular diffusivity, D, y may be expressed as,

$$y^2 \approx 2Dt$$

The average distance traveled by the fuel molecules in the radial direction will be of the order of the burner exit radius, R. Thus, the diffusion time may be estimated as,

$$R^2 \approx 2Dt \Rightarrow t \approx R^2/2D \tag{3.7}$$

If v is the fuel jet velocity at the nozzle exit, the time taken for the fuel molecule to reach the flame height, L_f, is given by $t = L_f/v$. This time should be same as that required for a fuel molecule to diffuse in the radial direction. Therefore, equating the convection and diffusion timescales, the flame length can be expressed as,

$$L_f \approx \frac{vR^2}{2D} \tag{3.8}$$

The fuel volume flow rate, Q_F, is expressed as $\pi R^2 v$. Using this, the expression for the flame length may be written as,

$$L_f \approx \frac{Q_F}{2\pi D} \tag{3.9}$$

It is clear that at a given fuel flow rate, the flame height is independent of the burner diameter and depends on Q_F, which may be obtained due to different combinations of burner diameter and jet velocity. Since the diffusion coefficient, D, is inversely proportional to pressure, the flame height is also inversely proportional to pressure at a given mass flow rate. As the volume flow rate is increased, the flame height will increase in the laminar regime, as demonstrated in Fig. 3.15.

For turbulent flows, the molecular diffusivity, D, may be replaced by turbulent mass diffusivity, which is expected to be of same order as that of turbulent eddy viscosity, v_t. Further, the eddy viscosity may be expressed as the product of turbulent mixing length, l_m, and turbulent intensity, v'_{rms}. Using these, the turbulent jet diffusion flame height may be written as,

$$L_{f,t} \approx \frac{vR^2}{v_t} \approx \frac{vR^2}{l_m v'_{rms}} \tag{3.10}$$

Further, it can be shown that the turbulent mixing length can be of the order of the jet radius (like an integral scale) and the maximum fluctuating component will be of the order of the jet velocity itself. Using these arguments, the turbulent flame height may be expressed as,

$$L_{f,t} \approx \frac{vR^2}{Rv} \approx R \tag{3.11}$$

This shows that the turbulent jet flame length depends on the port diameter alone, as mentioned earlier.

The variation of flame height as a function of jet velocity is shown in Fig. 3.18. As illustrated through photographs in Figs. 3.15 and 3.17, the length of jet diffusion flames in the laminar regime increases with fuel flow rate and in the turbulent regime, it remains almost a constant. Figure 3.18 also shows the envelope of the locations where flame oscillations start in the flame surface. It is clear that flame oscillations start from the tip and propagate upstream, and in the turbulent regime, the entire flame becomes oscillatory.

Based on the jet velocity, the jet diffusion flame may be in momentum-controlled regime or in buoyancy-controlled regime. The flame Froude number, Fr_f, defined as the ratio of the initial jet momentum to the buoyant force experienced by the flame is used to characterize the regime. Initial jet momentum is a function of jet exit velocity,

Fig. 3.18 Variation of height of jet diffusion flames as a function of fuel jet velocity for a given burner port diameter

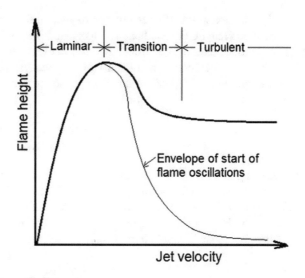

and the buoyant force is a function of the flame height itself. If flame Froude number is much greater than unity, $Fr_f \gg 1$, then the flame is momentum-controlled, since the jet momentum is higher than the buoyant force experienced by the flame. When $Fr_f \approx 1$, both jet momentum and buoyant force control the flame. On the other hand, when $Fr_f \ll 1$, then the flame is buoyancy-controlled.

Theoretical analysis of jet diffusion flames has been carried out by several researchers; first of them has been by Burke and Schumann, followed by Roper who proposed several correlations for flame height. Based on the experimental results for circular ports, the flame height in the laminar regime, whether it is in momentum- or buoyancy-controlled regime, may be expressed as,

$$L_f = 1330 \frac{Q_F(T_\infty/T_F)}{\ln(1 + 1/S)},$$

where S is the molar stoichiometric oxidizer to fuel ratio, T_∞ is the oxidizer stream temperature, T_F is the fuel stream temperature and Q_F is the volumetric flow rate of the fuel.

Laminar jet diffusion flame heights depend on the type of fuel, diluents added to the fuel stream, amount of air added to the fuel stream, called the primary air and so on. For a hydrocarbon fuel, C_xH_y, the molar stoichiometric air-to-fuel ratio, S, may be expressed as,

$$S = \frac{x + y/4}{X_{O_2}}.$$

The flame length increases as H/C ratio of the fuel decreases. When the mole fraction of the oxygen in the oxidizer increases, the flame length decreases. Even a small reduction in the mole fraction of oxygen in the air results in notable increase in the flame length. For methane jet emerging into a pure oxygen environment, the flame length is found to be around one-fourth of its value in an air environment.

When the fuel stream is diluted with an inert gas such as nitrogen, the molar stoichiometric air–fuel ratio is expressed using the mole fraction of the diluent, X_{dil}, as,

$$S = \frac{x + y/4}{[1/(1 - X_{dil})]X_{O_2}}.$$

It should be noted that as the mole fraction of the diluent increases, the flame length decreases. When primary air is added to the fuel, the flame length decreases significantly. In sooty fuels, this also reduces the net soot emission. If X_{pri} is the fraction of the stoichiometric air supplied as the primary air and S_{fuel} is the value of S when fuel alone is supplied to the burner, then the modified value of S is given as,

$$S = \frac{1 - X_{pri}}{X_{pri} + (1/S_{fuel})}.$$

For a turbulent flame, Guenther has reported a semiempirical correlation for the flame height ($L_{f,t}$), as,

$$\frac{L_{f,t}}{d} = 6(s + 1)\left(\frac{\rho_e}{\rho_F}\right)^{1/2}.$$

Here, d is the diameter of the fuel port, s is mass-based stoichiometric air–fuel ratio, ρ_e is fuel gas density and ρ_F is mean flame density, which is the mixture density calculated at an average temperature of around 1400 °C. Values for $L_{f,t}/d$ for various fuels calculated using this correlation are listed in Table 3.3.

Table 3.3 Typical values of turbulent jet flame lengths for various fuels obtained from Guenther's correlation

Fuel	$L_{f,t}/d$
Methane	200
Carbon monoxide	76
Acetylene	188
Propane	296
Hydrogen	147

3.2.3.1 Simplified Analysis of Diffusion Flames

Analysis of diffusion flames may be carried out using a simplified approach suggested by Spalding, called the Simple Chemical Reacting System (SCRS). The key assumptions are as follows:

(1) Even though several major and minor species contribute to the chemical reactions, the concentrations of the minor species, as the name suggests, are much less than those of the major species. Therefore, among the products, only major species are considered.

(2) Chemical reactions are much faster than the transport processes. Therefore, a global single-step reaction, as listed in Eq. (3.2), may be employed.

In a diffusion flame, where fuel and oxidizer streams are separate, Spalding defined a quantity called *conserved scalar*, which has no source term in its conservation equation. Consider a flow process in which f kg of fuel flows into the control volume through one inlet, $(1 - f)$ kg of oxidizer flows in through another inlet and 1 kg of product flows out through the exit. If E is a property associated with the flow, then for the mixing of fuel and the oxidizer stream, which yields a product stream to exit the control volume, the following relation holds good:

$$f E_F + (1 - f) E_A = E_M,$$

where E_F is the property of the fuel stream, E_A is that of the oxidizer or air stream and E_M is that of the mixture, which on completion of the chemical reaction will be the property of the product mixture. From this, f can be expressed as,

$$f = (E_M - E_A)/(E_F - E_A).$$

Here, f is called the *mixture fraction,* and in general, the property, E, will be a conserved scalar. Conserved scalars are obtained from any two regular flow variables such as mass fractions of fuel, oxidizer and products. Mixture enthalpy, defined in a specific manner, will also be a conserved scalar. For example, consider the global reaction where 1 kg of fuel and s kg of oxygen are consumed to form $(1 + s)$ kg of products. Then, a scalar variable involving mass fractions of fuel (Y_F) and oxygen (Y_{Ox}) may be defined such that the conservation equation involving that scalar variable will have no source terms. It can be inferred that for 1 kg of fuel consumed, s kg of oxidizer is consumed, and a scalar variable can be defined as,

$$E = Y_F - Y_{Ox}/s.$$

The variable E will not have any source term because,

$$S_E = \dot{m}_F''' - \dot{m}_{ox}'''/s = 0.$$

Using this scalar variable, the mixture fraction may be defined as,

$$f = \frac{[Y_F - Y_{Ox}/s]_M - [Y_F - Y_{Ox}/s]_A}{[Y_F - Y_{Ox}/s]_F - [Y_F - Y_{Ox}/s]_A}$$

It is understood that Y_{Ox} will be zero in the fuel stream and Y_F will be zero in the oxidizer stream. Also, Y_F in the fuel stream will be unity. Based on this, f may be written as,

$$f = \frac{[Y_F - Y_{Ox}/s]_M + [Y_{Ox}/s]_A}{1 + [Y_{Ox}/s]_A}$$

If the chemical reaction gets completed within the chamber, either fuel or oxygen gets consumed based on the value of f and will not be present in the product (M) stream. If f is equal to the stoichiometric value, then both fuel and oxygen will be consumed. Thus, f can have three possible values:

$$f < f_{stoich} : f = \frac{[-Y_{Ox}/s]_M + [Y_{Ox}/s]_A}{1 + [Y_{Ox}/s]_A}$$

$$f > f_{stoich} : f = \frac{[Y_F]_M + [Y_{Ox}/s]_A}{1 + [Y_{Ox}/s]_A}$$

$$f = f_{stoich} : f = \frac{[Y_{Ox}/s]_A}{1 + [Y_{Ox}/s]_A}$$

The extent of a diffusion flame can be depicted by the contour, $f = f_{stoich}$. Conserved scalars may also be defined using the mass fraction of the products, such as $Y_F + Y_P/(1 + s)$ and $Y_{Ox}/s + Y_P/(1 + s)$. In all such cases, the source term will be zero.

The mixture enthalpy is defined as,

$$h = c_p(T - T_{ref}) + \sum_j Y_j h_{j,T_{ref}}.$$

Here, c_p is the mixture specific heat, T_{ref} is the reference temperature (298 K) and $h_{j,T_{ref}}$ is the enthalpy of jth species at T_{ref}. Noting that the difference between enthalpy of reactant and that of the product should be the heat of combustion, Δh_C, h may be conveniently written as,

$$h = c_p(T - T_{ref}) + Y_F \Delta h_C$$
$$\text{Or } h = c_p(T - T_{ref}) + Y_{Ox} \Delta h_C/s$$

Since h defined in this manner is a conserved scalar, the mixture fraction, f, can be written in terms of h as follows:

$$f = \frac{c_p(T_M - T_{\text{ref}}) + Y_{F,M}\Delta h_C - c_p(T_\infty - T_{\text{ref}})}{\Delta h_C + c_p(T_F - T_\infty) - c_p(T_\infty - T_{\text{ref}})}.$$

Here, T_F is the temperature of the fuel stream, T_M is the product temperature, $Y_{F,M}$ is the mass fraction of the fuel in the product stream and T_∞ is the temperature of the oxidizer stream. Conveniently, the mixture fraction can be solved without involving any nonlinear source terms, and the original variables, such as mass fractions of the species and temperature, may be obtained from the mixture fraction field. For instance, at the flame sheet, the mass fractions of fuel and oxygen are obtained as follows:

When $f < f_{\text{stoich}}$, $Y_F = 0$, $Y_{Qx} = Y_{OX_x}(f_{\text{stoich}} - f)/f_{\text{stoich}}$.
When $f > f_{\text{stoich}}$, $Y_{Qx} = 0$, $Y_F = (f - f_{\text{stoich}})/(1 - f_{\text{stoich}})$.

At any f, the mass fraction of the inert species is obtained as,

$$Y_{\text{inert}} = Y_{\text{inert},A}(1 - f),$$

and the mass fraction of the product is obtained as,

$$Y_p = 1 - Y_F - Y_{Qx} - Y_{\text{inert}}.$$

Similarly, by solving the scalar transport equation for h^*, temperature distribution may be determined. At the flame sheet, the temperature may be evaluated as,

$$T = T(f) = f_{\text{stoich}}[(\Delta h_c/c_n) + T_F - T_\infty] + T_\infty$$

In this simplified analysis, temperature and mass fraction profiles vary linearly inside and outside the flame sheet. This is illustrated in Fig. 3.19.

3.3 Flames from Condensed Fuels

Condensed fuels such as solid and liquid fuels have to be gasified before they can take part in the gas-phase oxidation reactions. Heat is supplied to the surface of the condensed fuel, and as a result, gases or vapors are evolved. Therefore, rate of heat transfer to the surface of the condensed fuel becomes an important factor. To enable higher rates of heat and mass transfers, condensed fuels are disintegrated into small particles; solid fuels are pulverized to form small particles, and liquid fuels are atomized to generate small droplets, such that the ratio of the surface area to volume increases. Burning of these particles and droplets will be heterogeneous in nature due to the presence of multiple phases. Before analyzing how these particles and droplets burn in a combustion chamber, gasification and burning characteristics of individual particles or droplets should be understood. Toward this objective, the

Fig. 3.19 Diffusion flame
structure obtained using
SCRS

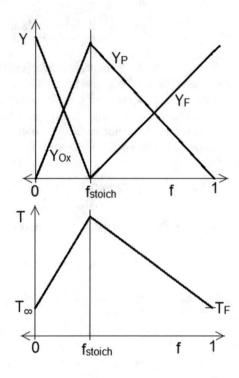

evaporation and burning characteristics of a liquid fuel droplet and followed by that
the combustion of an isolated carbon particle are discussed next.

3.3.1 Evaporation and Burning of Isolated Fuel Droplets

3.3.1.1 Droplet Evaporation

A disintegrated liquid jet produces small sheets and ligaments, which due to surface
tension force form small droplets. Under low to moderate pressure conditions, a
droplet evaporates or burns in a quasi-steady manner. This means that even though
the surface area decreases with time as a result of phase change, the rate at which it
decreases is a constant. Heat from the ambient supplies energy to vaporize the liquid,
and the vapor thus produced diffuses from the droplet surface into the ambient. This
process continues until the droplet completely evaporates.

Consider a liquid fuel droplet having a radius of r_s, present in a quiescent environment at a temperature T_∞ as shown in Fig. 3.20. Let T_s be the surface temperature of
the droplet. Thermodynamic equilibrium prevails on the droplet surface. This means
that for the given surface temperature, there is an associated saturation pressure,
p_{sat}. This saturation pressure is the same as the partial pressure of the fuel vapor

Fig. 3.20 Schematic of
vaporization of a spherical
droplet in a quiescent
environment

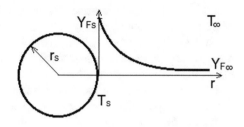

at the liquid–vapor interface. The partial pressure of the vapor at the interface may
be calculated from the surface temperature of the droplet using a typical thermodynamic relationship such as Clausius–Clapeyron equation or Wagner equation. Since
the partial pressure is related to the mole fraction of the fuel, the mass fraction of the
fuel at the interface, Y_{Fs}, may be determined.

Initially, the heat transfer from the ambient is utilized for two purposes: to raise
the temperature of the liquid and to supply the latent heat of vaporization. During
this period, the droplet diameter can increase as a result of volumetric expansion,
which is dependent on the liquid-phase properties of the fuel. This aspect is shown as
phase 1 in Fig. 3.21, where the temporal variation of square of the droplet diameter
(representing the surface area) has been schematically presented. After reaching a
maximum diameter, the evaporation process starts, in which part of the heat is used
as the latent heat and the remainder is used for heating the liquid phase. This is
shown as phase 2 in Fig. 3.21. However, after a certain time, the droplet surface
attains a temperature called equilibrium temperature, which is a function of the
ambient temperature and the boiling point of the liquid fuel. At this condition, the heat
transfer to the droplet surface exactly equals the product of the vaporization rate and
the latent heat of vaporization. The surface regression of the droplet shows a linearly
decreasing trend, as shown in phase 3 in Fig. 3.21. For equilibrium vaporization,
the surface temperature cannot exceed the boiling point of the fuel, even when the
ambient temperature is much higher than the boiling point. In fact, the equilibrium
temperature is a few degrees less than the boiling point due to a process called
evaporative cooling. In general, the time taken for phases 1 and 2 is no more than 10
to 15% of the total evaporation period. Thus, these phases can be neglected while

Fig. 3.21 Schematic
representation of surface
regression of a typical fuel
droplet exhibiting different
phases (1, 2 and 3) during its
vaporization

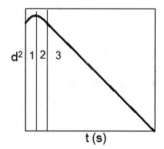

calculating the evaporation rate. The droplet lifetime is the time when the diameter of the droplet becomes zero.

To determine the evaporation rate during equilibrium evaporation, a simplified mass transfer analysis may be carried out. Here, the properties such as density and mass diffusivity are taken to be constants. This is acceptable since the attention is focused in the vicinity of the droplet surface. The mass escaping from the droplet surface to the gas phase is transported by convection and diffusion.

If ρ_F is the vapor density of the fuel and v_F is the velocity normal to the surface, then the mass flux of the vapor at the droplet surface is given as $\rho_F v_F$. Using Fick's law of ordinary diffusion, the mass transport may be expressed as,

$$\rho_F v_F = \rho V Y_F - \rho D_{FA} \frac{dY_F}{dr}. \tag{3.12}$$

Here, V is the bulk or mixture velocity, which is equal to the fuel velocity, since the ambient fluid is stagnant. D_{FA} is the binary diffusivity, which dictates the diffusion of fuel into the ambient fluid, ρ is the mixture density and Y_F is the mass fraction of the fuel. It is clear that the mass flux by diffusion is dependent on the gradient of the fuel mass fraction and takes place in the direction of decreasing concentration of fuel. Equation (3.12) may be rewritten, noting that $\rho Y_F = \rho_F$, as,

$$\rho_F v_F = \rho_F v_F - \rho D_{FA} \frac{dY_F}{dr} \Rightarrow \rho_F v_F = -\frac{\rho D_{FA}}{1 - Y_F} \frac{dY_F}{dr}. \tag{3.13}$$

If $\rho_F v_F$ is multiplied with $4\pi r^2$, the evaporation rate ($\dot{m}$) may be obtained. Equation (3.13) can be integrated, and boundary condition on the droplet surface, Y_F ($r = r_s) = Y_{Fs}$, may be applied to determine the variation of fuel mass fraction along the radial direction. This leads to,

$$Y_F(r) = 1 - \frac{(1 - Y_{F,s}) \exp[-\dot{m}/(4\pi \rho D_{FB} r)]}{\exp[-\dot{m}/(4\pi \rho D_{FB} r_s)]}. \tag{3.14}$$

By using the condition Y_F ($r \to \infty) = Y_{F\infty}$, in Eq. (3.14), the evaporation rate ($\dot{m}$) may be determined as,

$$\dot{m} = 4\pi r_s^2 \rho_F v_F = 4\pi r_s \rho D_{FA} \ln\left(\frac{1 - Y_{F\infty}}{1 - Y_{Fs}}\right). \tag{3.15}$$

The evaporation rate is dependent on radius of the droplet, mixture density and mass diffusivity. Spalding defined a non-dimensional number called mass transfer number, given as,

$$B_Y = (Y_{FS} - Y_{F\infty})/(1 - Y_{Fs}).$$

The quantity whose natural logarithm is used in Eq. (3.15) may be then expressed as $1 + B_Y$. Using this, Eq. (3.15) may be rewritten as,

$$\dot{m} = 4\pi r_s \rho D_{FA} \ln(1 + B_Y).$$ (3.16)

The mass transfer number indicates how fast the evaporation takes place. Due to evaporation, the rate at which the mass of the droplet, m_d, decreases may be expressed as,

$$\frac{dm_d}{dt} = -\dot{m}.$$ (3.17)

The mass of the droplet is expressed as the product of the liquid density and its volume,

$$m_d = \rho_\ell \frac{\pi d^3}{6}$$

Writing the droplet radius in terms of its diameter, d, Eq. (3.17) may be written as,

$$\frac{d(d)}{dt} = \frac{4\rho D_{FA}}{\rho_\ell d} \ln(1 + B_Y) \Rightarrow \frac{d(d^2)}{dt} = -\frac{8\rho D_{FA}}{\rho_\ell} \ln(1 + B_Y).$$ (3.18)

The rate of change of diameter square is the surface regression rate of the droplet. The term in the RHS of Eq. (3.18) is called the *evaporation rate constant*, K_v.

$$K_v = \frac{8\rho D_{FA}}{\rho_\ell} \ln(1 + B_Y).$$ (3.19)

Using the initial condition that at time $t = 0$, $d = d_0$, the initial droplet diameter, integration of Eq. (3.18) yields the famous d^2-law: $d^2 = (d_0)^2 - K_v t$. It is important to note that the evaporation constant is independent of the droplet diameter. From droplet evaporation experiments, instantaneous droplet diameters may be recorded using optical methods. When the square of instantaneous droplet diameter is plotted as a function of time, a straight line with a negative slope is obtained. The slope of this line yields the value of the evaporation constant.

Under normal gravity conditions and/or if there is a relative velocity between the droplet and the ambient fluid, the evaporation rate is enhanced. This is due to the enhancement in the gradient of the fuel mass fraction at the droplet surface, which is caused by convective (natural or forced or both) transport of the fuel vapor away from the droplet surface. Also, internal flow field is created within the droplet due to the shear stress induced by the external gas-phase flow. Marangoni convection, caused due to surface tension gradients, also contributed to the shear stress at the droplet surface. Typical streamlines outside and within the droplet are shown in Fig. 3.22.

Fig. 3.22 Streamlines in gas phases and liquid phases for a droplet evaporating in the presence of convection

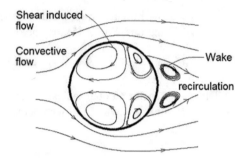

Even under a convective environment, the evaporation may be quasi-steady for a major portion of the droplet lifetime. Under convective conditions, the Nusselt number (Nu) is introduced into Eqs. (3.16 and 3.19) in order to account for convective heat transfer. Nusselt number is calculated as a function of Reynolds number and Prandtl number for forced convection, and it is evaluated as a function of Grashoff number and Prandtl number in the case of natural convection. The expressions for evaporation rate constant for droplet evaporating under convective condition are expressed as,

$$K_v = \frac{4 \mathrm{Nu} \rho D_{\mathrm{FA}}}{\rho_\ell} \ln(1 + B_Y).$$

In general, the Nusselt number may be written as $\mathrm{Nu} = 2 + f(\mathrm{Re}, \mathrm{Pr})$ for forced convective environment or $\mathrm{Nu} = 2 + f(\mathrm{Gr}, \mathrm{Pr})$ for natural convection. Therefore, under zero convection, either $\mathrm{Re} = 0$ or $\mathrm{Gr} = 0$ and the limiting value of Nu is 2, which gives back Eq. (3.19) for droplet evaporation in the absence of convection.

In the development above, the evaporation rate has been obtained by considering the mass transfer analysis. The same may be estimated by using energy balance. The overall energy balance equation may be written as,

$$\dot{m}'' c_{\mathrm{pg}} \frac{dT}{dr} = \frac{1}{r^2} \frac{d}{dr} \left(r^2 k \frac{dT}{dr} \right) \Rightarrow \frac{\dot{m} c_{\mathrm{pg}}}{4 \pi k} \frac{dT}{dr} = \frac{d}{dr} \left(r^2 \frac{dT}{dr} \right). \tag{3.20}$$

Here, c_{pg} is the gas-phase specific heat and k is thermal conductivity of the vapor mixture. Equation (3.20) may be solved by applying the following boundary conditions; $T(r = r_s) = T_s$ and $T(r \to \infty) = T_\infty$. The temperature distribution may be expressed as,

$$T(r) = \frac{(T_\infty - T_s) \exp(-Z\dot{m}/r) - T_\infty \exp(-Z\dot{m}/r_s) + T_s}{1 - \exp[-Z\dot{m}/r_s]}, \tag{3.21}$$

where $Z = c_{\mathrm{pg}}/(4\pi k)$. For equilibrium evaporation, the boundary condition at the droplet surface is given by,

$$\dot{m}h_{fg} = 4\pi kr_s^2 \left[\frac{dT}{dr} \right]_{r_s}. \tag{3.22}$$

The boundary derivative may be evaluated by differentiating Eq. (3.21) with respect to r and setting $r = r_s$. Substitution of this into Eq. (3.22) leads to the following expression for the evaporation rate:

$$\dot{m} = \frac{4\pi kr_s}{c_{pg}} \ln \left(1 + \frac{c_{pg}(T_\infty - T_s)}{h_{fg}} \right). \tag{3.23}$$

Spalding defined another transfer number, B_T, which is based on heat transfer, from Eq. (3.23), as follows:

$$B_T = \frac{c_{pg}(T_\infty - T_s)}{h_{fg}}. \tag{3.24}$$

In equilibrium evaporation, the mass evaporation rate obtained by mass transfer analysis, given by Eq. (3.16), and that obtained by Eq. (3.24) are the same. That is,

$$\dot{m} = \frac{4\pi kr_s}{c_{pg}} \ln(1 + B_T) = 4\pi r_s \rho D_{FA} \ln(1 + B_Y)$$

This implies that the transfer numbers based on mass transfer and heat transfer may be related as,

$$1 + B_Y = (1 + B_T)^{Le}. \tag{3.25}$$

Here, Le $= k/(\rho D_{FA} c_{pg})$ is a dimensionless number called the Lewis number. When the molecular weight of the fuel is almost the same as that of the ambient fluid, the value of Le is close to unity. If the molecular weight of the vapor is higher, then Le is greater than unity. The mass fraction of the fuel at the droplet surface may be evaluated from Eq. (3.25) as,

$$Y_{Fs} = 1 - \frac{1 - Y_{F\infty}}{\left[1 + \frac{c_{pg}(T_\infty - T_s)}{h_{fg}} \right]^{Le}}. \tag{3.26}$$

It is clear that when T_∞ tends to infinity or h_{fg} tends to zero, Y_{Fs} tends to unity. Since the mass fraction of fuel vapor at the droplet surface is a function of T_s, the values of T_s and Y_{Fs} may be evaluated by simultaneously solving a thermodynamic relation such as Clausius–Clapeyron equation and Eq. (3.26). This is illustrated in Fig. 3.23.

As an alternative, the equilibrium temperature, T_s, for various values of the ambient temperature may be determined using Eq. (3.25), using the following procedure:

Fig. 3.23 Equilibrium
surface temperatures
[intersections of curves of a
thermodynamic relation and
those obtained by solving
Eq. (3.26)] at different
ambient temperatures

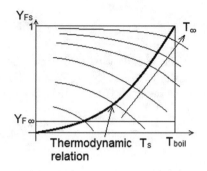

1. First a value of T_s is assumed.
2. Using the thermodynamic relation, the saturation pressure at T_s is evaluated.
3. The mole and mass fractions of the fuel at the droplet surface are calculated.
4. The values of B_T and B_Y are calculated.
5. Substitute these values in Eq. (3.25), and determine if the equation is satisfied.
6. If Eq. (3.25) is not satisfied, the value of T_s is adjusted and steps 2 to 5 are
   repeated until Eq. (3.25) is satisfied.

In the case of a droplet constituted by liquid fuel blends, the vaporization characteristics are significantly different. For instance, if a droplet has hexane and decane
mixed in some proportion by volume, the surface regression curve has a higher slope
in the beginning indicating the preferential vaporization of hexane, the more volatile
component. Upon completion of evaporation of hexane, surface regression curve
exhibits a lesser slope, indicating the evaporation of less volatile component. This is
schematically illustrated in Fig. 3.24.

Fig. 3.24 Profile of surface
regression of a
multi-component droplet
showing preferential
vaporization of the higher
volatile component

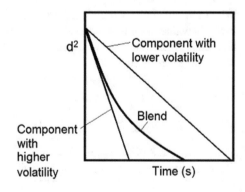

3.3.1.2 Droplet Combustion

When an ignition source is provided to an evaporating droplet, a diffusion flame is established around the droplet, at a certain radius r_f. This is schematically depicted in Fig. 3.25. As in any diffusion flame, fuel from the droplet surface ($r = r_s$) reaches the flame zone by convection and diffusion, where it is consumed. Similarly, oxygen from the ambient is transported toward the flame zone and it is consumed almost completely in the flame zone. In general, the transport processes, and more specifically the evaporation process in this heterogeneous case, are much slower than the chemical kinetics. Therefore, this problem may be analyzed by considering only the transport processes involved in mass and energy conservation equations, as done in SCRS. Further, when a flame surrounds the droplet, it is similar to a droplet evaporating in a very high ambient temperature environment. However, in the case of droplet evaporation, the ambient fluid (oxidizer) reaches the droplet surface due to diffusion. In the case of droplet combustion, in contrast, oxygen is consumed in the flame zone and the products of combustion, formed in the flame zone, diffuse toward the droplet surface. Under low to moderate pressure conditions, equilibrium vaporization and subsequent combustion of the fuel vapors are possible. Even when a flame surrounds the droplet, the surface temperature is a few degrees less than the boiling point of the liquid.

Consider a conserved property, ϕ. Its conservation in terms of the net flux of that property from the droplet surface $r = r_s$ to any radius r, which occurs due to convection and diffusion, can be written as,

$$\left(\rho v\phi - \rho D\frac{d\phi}{dr}\right)r^2 = \left(\rho_s v_s \phi_s - \rho_s D_s\left[\frac{d\phi}{dr}\right]_s\right)r_s^2. \qquad (3.27)$$

Fig. 3.25 Schematic representation of spherically symmetric droplet combustion and the profiles of temperature and species mass fractions

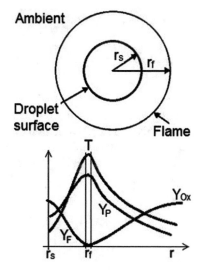

If ρD is assumed to be constant, Eq. (3.27) can be integrated and applying the boundary condition for $r \to \infty$, the evaporation/burning rate may be expressed as,

$$\rho_s v_s = \frac{\rho D}{r_s} \ln\left[1 + \frac{\rho_s v_s (\phi_s - \phi_\infty)}{-\rho D\left(\frac{d\phi}{dr}\right)_s}\right]. \tag{3.28}$$

If $\phi = Y_F - sY_{Ox}$, then $\phi_\infty = -Y_{Ox\infty}/s$, because there is no fuel in the ambient across the flame, and $\phi_s = Y_{Fs}$, because there is no oxygen at the droplet surface. Further, by Fick's law,

$$-\rho D\left(\frac{d\phi}{dr}\right)_s = -\rho D\left(\frac{dY_F}{dr}\right)_s = \rho_s v_s (1 - Y_{Fs}). \tag{3.29}$$

Using these in Eq. (3.28), the burning rate per unit area may be expressed as,

$$\rho_s v_s = \frac{\rho D}{r_s} \ln\left[1 + \frac{\rho_s v_s (Y_{Fs} + Y_{Ox\infty}/s)}{\rho_s v_s (1 - Y_{Fs})}\right] = \frac{\rho D}{r_s} \ln\left[1 + \frac{(Y_{Fs} + Y_{Ox\infty}/s)}{(1 - Y_{Fs})}\right]. \tag{3.30}$$

If $\phi = c_p(T - T_\infty) + Y_F \Delta h_c$, then $\phi_\infty = 0$ and $\phi_s = c_p(T_s - T_\infty) + Y_{Fs} \Delta h_c$. Further, assuming that all properties are constant and $\mathrm{Le} = 1$, that is, $k/c_p = \rho D$,

$$\begin{aligned} -\rho D\left(\frac{d\phi}{dr}\right)_s &= -\rho D\left(\frac{d}{dr}(c_p(T - T_\infty) + Y_F \Delta h_c)\right)_s \\ &= -k\left(\frac{dT}{dr}\right)_s - \Delta h_c \rho D\left(\frac{dY_F}{dr}\right)_s. \end{aligned} \tag{3.31}$$

Using Eq. (3.29), Eq. (3.31) may be expressed as,

$$-\rho D\left(\frac{d\phi}{dr}\right)_s == -k\left(\frac{dT}{dr}\right)_s + \Delta h_c \rho_s v_s (1 - Y_{Fs}) = q_s + \Delta h_c \rho_s v_s (1 - Y_{Fs}). \tag{3.32}$$

Here, q_s is the heat flux to the droplet surface from the ambient. As discussed earlier, this heat flux is used initially for heating the droplet until it reaches an equilibrium temperature and then for supplying the latent heat of vaporization. Using the definitions of ϕ_s, ϕ_∞ and Eq. (3.32) in Eq. (3.28), the burning rate per unit area may be written as,

$$\rho_s v_s = \frac{\rho D}{r_s} \ln\left[1 + \frac{\rho_s v_s (c_p(T_s - T_\infty) + Y_{Fs} \Delta h_c)}{q_s + \Delta h_c \rho_s v_s (1 - Y_{Fs})}\right]$$

$$= \frac{\rho D}{r_s} \ln \left[1 + \frac{(c_p(T_s - T_\infty) + Y_{Fs}\Delta h_c)}{(q_s/p_s v_s) + \Delta h_c(1 - Y_{Fs})} \right] \tag{3.33}$$

Similarly, if $\phi = c_p(T - T_\infty) + Y_{Ox}\Delta h_c/s$, then it may be shown that,

$$\rho_s v_s = \frac{\rho D}{r_s} \ln \left[1 + \frac{(c_p(T_s - T_\infty) + Y_{Ox\infty}\Delta h_c/s)}{-(q_s/p_s v_s)} \right]. \tag{3.34}$$

In general, T_∞ and $Y_{Ox\infty}$ are known quantities. Even though the conditions at droplet surface are somewhat uncertain, there is a definite relationship between T_s and Y_{Fs}. Further, T_s cannot exceed the value of the boiling point of the liquid. Therefore, the equilibrium surface temperature is determined by equating Eqs. (3.30 and 3.34):

$$\frac{(c_p(T_s - T_\infty) + Y_{Ox\infty}\Delta h_c/s)}{-(q_s/p_s v_s)} = \frac{(Y_{Fs} + Y_{Ox\infty}/s)}{(1 - Y_{Fs})}. \tag{3.35}$$

The quantity, $q_s/(\rho_s v_s)$ is equal to $h_{fg} + c_{liq}(T_{boil} - T_0)$, where h_{fg} is latent heat of vaporization, c_{liq} is specific heat capacity of the liquid fuel, T_{boil} is its boiling point and T_0 is its initial temperature. In a combustion scenario, since T_∞ is very high, close to the flame temperature, it may be shown that the surface temperature is only a few degrees less than the boiling point and may be assumed to be same as the boiling point itself. Therefore, a useful formula for the burning rate of a spherically symmetric droplet is,

$$\rho_s v_s = \frac{\rho D}{r_s} \ln \left[1 + \frac{(c_p(T_\infty - T_{boil}) - Y_{Ox\infty}\Delta h_c/s)}{h_{fg} + c_{liq}(T_{boil} - T_0)} \right]. \tag{3.36}$$

When the equilibrium temperature is reached, the value of $q_s/(\rho_s v_s)$ will be equal to h_{fg} only. In summary, in a combustion environment, transfer numbers may be defined in several ways:

$$B = \frac{(Y_{Fs} + Y_{Ox\infty}/s)}{(1 - Y_{Fs})} \text{ or } \frac{(c_p(T_s - T_\infty) + Y_{Fs}\Delta h_c)}{(q_s/p_s v_s) + \Delta h_c(1 - Y_{Fs})}$$

$$\text{or } \frac{(c_p(T_s - T_\infty) + Y_{Ox\infty}\Delta h_c/s)}{-(q_s/p_s v_s)}.$$

Here, conveniently, T_s may be replaced with T_{boil} and $q_s/(\rho_s v_s)$ may be replaced by h_{fg} for quasi-steady burning.

Droplet combustion in a convective environment and combustion of multi-component fuel droplets have characteristics similar to corresponding vaporizing droplets. Features such as liquid-phase flow induced by shear of the external flow as well as Marangoni convection and preferential vaporization and combustion of higher volatile component are observed. In case of alcohol fuels such as methanol and ethanol, the water vapor which is formed in the flame zone diffuses toward the droplet surface and partially condenses on the surface due to the low boiling point of the alcohol. Since water and alcohol are polar molecules, the alcohol droplet absorbs the condensed water. This causes a variation in the mass burning rate and can also cause extinction of the flame.

In a spray environment, droplets of several sizes are injected with different velocities into the combustion chamber. Each droplet evaporates at a rate that is dependent on its diameter and relative velocity in addition to the ambient temperature. The fuel vapors mix with the oxygen present in the chamber. If the temperature exceeds the auto-ignition temperature, the mixture ignites. A flame can thus initiate in the wake region of the moving droplet and eventually envelop it. The droplet velocity decreases due to drag, and there is a certain penetration distance associated with the spray as a whole. The more complex part in a spray combustion environment is the interference effects between the neighboring droplets. While the heat transferred from the flame around a droplet may enhance the vaporization rate of the neighboring droplet, there is competition for oxygen between the droplets. Therefore, the analysis of entire spray combustion is not straightforward and requires a statistical approach. Some aspects of liquid fuel combustion in practical combustors are discussed in a later chapter.

3.3.2 Combustion of a Carbon Particle

Solid fuels such as coal and biomass are called charring-type fuels, as they leave carbonaceous residue and ash when heated in an inert environment. The carbon content in these fuels varies based on the source from which they are obtained. The oxidation of carbon forms the rate-limiting step in the combustion of these fuels. As such, the combustion of coal and biomass is highly complex and heterogeneous in nature. For a practical analysis of the combustion of such fuels, it turns out that a thorough understanding of the combustion of a carbon particle is necessary. This is discussed next.

Combustion of carbon particle involves the following steps:

(1)   Diffusion of oxygen toward the particle surface.
(2)   Surface absorption of oxygen by the carbon.
(3)   Occurrence of surface reaction to form products embedded within the surface.
(4)   Products desorbed from the surface.
(5)   Diffusion of products toward the ambient.

Oxidation of carbon due to surface reactions increases the temperature of the particle. The particle becomes incandescent due to its high temperature. Consequently, loss of heat due to thermal radiation becomes significant and the surface temperature decreases and attains a value of approximately 900 °C. If the particle temperature is high enough, the carbon–oxygen reaction will occur at a very fast rate.

If the flow rate of air relative to the carbon particle is high enough, or if the particle size is high enough, then diffusion of oxygen into the carbon surface would occur at a faster rate. This would also increase the rate at which the product desorbs from the carbon surface and diffuses away from the surface. This regime is predominantly controlled by oxygen diffusion and is hence called the diffusion-controlled regime. This regime is dependent on particle size, its temperature and the surrounding flow field.

On the other hand, if the particle temperature is relatively low, particle is small and the relative flow rate of air is also less, then chemical kinetics would dictate the burning process. This regime is called kinetically controlled regime. The gradients in temperature and oxygen concentration are much smaller in this regime. The kinetics and, hence, the burning rate exponentially increase with an increase in the temperature.

The diffusion-controlled and kinetically controlled regimes are shown in Fig. 3.26. The typical variation of burning rate as a function of particle temperature is shown in Fig. 3.27. It is clear that temperature controls the burning rate in the kinetics regime and the particle size controls the burning rate in the diffusion regime.

Fig. 3.26 Radial profiles of oxygen and temperature in kinetically and diffusion-controlled regimes in carbon particle combustion

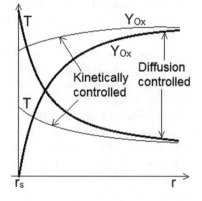

Fig. 3.27 Carbon burning rate as a function of temperature

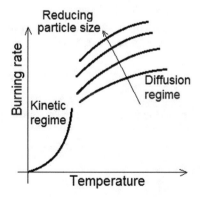

Fig. 3.28 Diffusion regime; profiles in one-film model

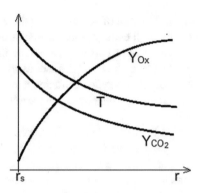

In general, experimental observations show that carbon oxidation is diffusion-controlled. This is because at low temperatures, the particle tends to loose heat and consequently extinction occurs. In the diffusion regime, carbon oxidation occurs in one of three ways. When sufficient oxygen diffuses to the particle surface and the particle temperature is also high enough, carbon dioxide is formed as a result of the surface reaction: $C + O_2 \rightarrow CO_2$, and CO_2 formed at the surface diffuses away toward the ambient. This is the concept used in the *one-film* model. The profiles of oxygen, CO_2 and the temperature are shown in Fig. 3.28. In the second case, if enough oxygen diffusion does not occur, then CO is formed through the surface reaction, $2C + O_2 \rightarrow 2CO$. The CO thus formed, based on the availability of oxygen, is converted to CO_2 or leaves as CO itself.

The third case is based on experimental observations. From experiments, since negligible amounts of CO_2 and O_2 are observed near the carbon surface, researchers hypothesized that this must be due to the occurrence of the reaction $C + CO_2 \rightarrow 2CO$, which is a *reduction reaction*, taking place at the particle surface. The CO thus formed at the carbon surface diffuses toward the ambient, where it mixes with the O_2 and forms CO_2, through the reaction $CO + 0.5O_2 \rightarrow CO_2$. Temperature attains its maximum value at the radial location where CO_2 is formed. Experimentally, a thin, non-luminous, bluish flame is usually seen to surround the dark carbon particle. This is the concept of *two-film* model, and the profiles of O_2, CO_2, CO and temperature are shown in Fig. 3.29.

In the more realistic two-film model, the stoichiometric relation at the carbon surface may be written as,

$$1\text{kg C} + v_s\text{ kg CO}_2 \rightarrow (1 + v_s)\text{kg CO},$$

and at the flame sheet the stoichiometric relation is expressed as,

$$1\text{kg CO} + v_f\text{ kg O}_2 \rightarrow (1 + v_f)\text{kg CO}_2.$$

Fig. 3.29 Diffusion regime;
profiles in two-film model

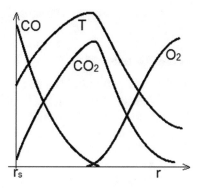

Applying Fick's law for CO_2 transport and using appropriate boundary conditions in the inner (between the carbon surface and the flame) and the outer (between the flame and the ambient) zones, two equations are formulated. Applying the same for the transport of the inert species, nitrogen, one more equation is obtained. However, there are five unknowns, namely burning rate of carbon, mass fractions of CO_2 at the flame and at the carbon surface, mass fraction of nitrogen at the carbon surface and the flame radius. At the flame sheet, $Y_{Nitrogen} = 1 - Y_{CO2}$. This forms the fourth equation, and one more equation is needed. The reduction reaction, $C + CO_2 \rightarrow 2CO$, has a rate that may be expressed as,

$$\omega_C = \dot{m}''_{C,s} = k_C MW_C [CO_{2,s}].$$

Here, $k_C = A \exp[-E_a/R_u T]$ is rate coefficient, E_a is activation energy, R_u is universal gas constant, ω_C is the carbon reaction rate in kg/m^2 s, $[CO_{2,s}]$ is the concentration of CO_2 at the carbon surface and MW_C is molecular weight of the carbon. Expressing the concentration in terms of mass fraction using the equation of state and multiplying by the surface area, the burning rate may be obtained as,

$$\dot{m}_C = 4\pi r_s^2 k_C \frac{p}{R_s T_s} Y_{CO_{2s}}.$$

Here, R_s is calculated using universal gas constant and molecular weights of C, mixture and CO_2. This kinetic equation forms the final equation required to solve the 5 variables in the two-film model. Finally, an expression for carbon burn rate may be obtained as,

$$\dot{m}_C = 4\pi r_s \rho D \ln\left(1 + \frac{2Y_{O_{2\infty}} - [(\nu_s - 1)/\nu_s]Y_{CO_{2s}}}{(\nu_s - 1) + [(\nu_s - 1)/\nu_s]Y_{CO_{2s}}}\right).$$

In reality, the coal combustion is much more complicated. The particle is usually non-spherical, porous and subjected to swelling as well as shrinkage. It also contains minerals such as ash and so on, resulting in a very complex heterogeneous combustion process. Coal combustion in several practical combustors is dealt with in a separate chapter.

Review Questions

1. What are the criteria required for the onset of piloted ignition?
2. When is a reactant mixture called flammable?
3. Define flammability limits.
4. Draw the typical structure of a premixed flame.
5. How an estimate of flame speed is obtained by Mallard and Le Chatelier?
6. List the parameters on which the flame speed would depend on.
7. Explain how the flame speed is affected by pressure.
8. Define various turbulent length scales.
9. Explain the different regimes of turbulent premixed flames.
10. Explain how flame speed is enhanced in the case of a wrinkled flame.
11. What are the stability problems observed in premixed flames?
12. How the quenching diameter is estimated?
13. List the differences between a premixed flame and a diffusion flame.
14. How the flame length varies with fuel flow rate in a jet diffusion flame?
15. What are the effects of addition of diluents and primary air to the fuel stream on jet flame length?
16. Draw the structure of jet diffusion flame at several axial locations from burner exit.
17. Use Roper's correlation to evaluate the length of methane jet diffusion flame and to calculate the flame length when hydrogen is added to methane in different volumetric proportions.
18. Comment on whether Roper's correlation can be used to predict the flame lengths of CO–hydrogen flames.
19. What is a conserved scalar?
20. Draw the structure of jet diffusion flame obtained by SCRS.
21. What is thermodynamic phase equilibrium?
22. What are the typical phases observed in droplet evaporation?
23. What is Fick's law of ordinary diffusion?
24. How is transfer number defined based on mass and heat transfer analyses?
25. What is d^2-law?
26. How the evaporation rate equation will change in convective conditions?
27. How is equilibrium surface temperature determined?
28. Comment on vaporization characteristics of multi-component fuel droplets.
29. Write the possible conserved scalar variables associated with droplet combustion.
30. Apply Spalding's SCRS analysis to one-film model for carbon combustion.

Exercise Problems

1. In a compartment of 2 × 3 × 5 m, air is present at 298 K and 1 bar. An acetylene cylinder kept in the room leaks at a rate of 1 L/m for an hour before it is emptied. Determine if the mixture formed is flammable.

2. A cylindrical tube burner of 60-mm diameter is supplied with a premixed mixture of butane and air. The volumetric flow rate of the mixture is 17.5 L/m, and the volumetric fraction of butane is 0.025. The laminar flame speed in cm/s may be estimated using the correlation, $S_L = 34 - 139(\phi - 1)^2$, where ϕ is the equivalence ratio. Determine the direction and velocity of the flame propagation.

3. Laminar flame speed of a reactant mixture of methane and air at 1 atm and 298 K is 0.4 m/s. Nitrogen in air is replaced by argon, such that the mole fractions of components in air are 0.21, 0.29 and 0.5, for oxygen, nitrogen and argon, respectively. For the same reactant temperature and pressure, what is the resultant laminar flame speed?

4. Consider a cylindrical burner of diameter 50 cm sustaining a premixed flame of propane–air. The fuel flow rate is 0.8 kg/s. The mean reactant velocity is 1 m/s, and mean flame temperature is 2000 K. If the turbulent intensity is 18% of the mean value and the length scale involved is 20% of the burner diameter, determine the possible regime for the turbulent flame.

5. Determine the length of an ethane diffusion flame in open air for a fuel velocity of 5 cm/s. The velocity profile issuing from the 10-mm-diameter port is uniform. Both the air and ethane are at 300 K and 1 atm.

6. Plot the variation of flame length when butane is supplied through a 10-mm-internal diameter cylinder burner at an average velocity of 10 cm/s for the cases with primary air varied as 0, 10, 20 and 40% of the stoichiometric value, (a) keeping the fuel flow rate constant and (b) keeping the reactant flow rate constant.

7. LPG (60% butane and 40% propane by volume) is supplied through a cylindrical burner of 10-mm internal diameter at a mass flow rate of 0.25 kg/hour. Determine the flame height.

8. Calculate the equilibrium surface temperature of a methanol droplet of 0.5-mm diameter evaporating in a nitrogen ambient at (a) 300 K, (b) 600 K and (c) 1800 K. Also, determine the mass evaporation rates for these cases. Gas-phase thermal conductivity, specific heat at constant pressure and mass diffusivity may be taken as 0.09 W/m K, 1320 J/kg K and 6.5 × 10⁻⁵ m²/s, respectively. Other properties may be taken from NIST database.

9. Calculate the mass evaporation rate of 1-mm-diameter droplet of 60% methanol and 40% water in a nitrogen environment at 400 K.

10. Estimate the mass burning rate of a n-heptane droplet of initial diameter of 1 mm burning in air at atmospheric pressure. Assume no heat is conducted into the interior of the liquid droplet and that the droplet temperature is equal to the boiling point minus 10 K. The ambient air is at 298 K. Values of specific heat at constant pressure and thermal conductivity may be taken as 4390 J/kg K and 0.111 W/m K, respectively. Other properties may be taken from NIST database.

Chapter 4
Burners for Gaseous Fuels

In the previous chapter, characteristics of premixed and non-premixed flames were presented. In general, premixed flames are seen to be shorter, cleaner and burn without much soot. By varying the composition of the reactant mixture, the flame temperature may be controlled. However, they are prone to instabilities such as lift-off and flashback due to the mismatches between reactant flow speed and laminar flame speed. Flashback is not only a stability problem but also a safety hazard. These issues restrict the useful operation range of a premixed flame burner. Moreover, under realistic operating conditions, the flames are very likely to be turbulent and turbulent premixed flames are more difficult to control owing to their increased complexity. Jet diffusion flames are more stable and consequently have a wider operation range and controllability. However, they tend to be quite long, soot formation is unavoidable, especially depending on the type of fuel used, and turbulent diffusion flames are highly noisy. Hence, it is sensible to combine the advantages of both these types of flames while designing practical gaseous fuel burners. The ways and means by which this can be achieved are discussed in this chapter.

4.1 Classifications

Burners for gaseous fuels may be broadly classified as *direct-fired* or *open flame* burners and *indirect* or *closed* burners. In direct-fired burners, different types of flames may be established in the combustion chamber. The hot gases transfer the heat to the required system and exit through an exhaust. On the other hand, in the indirect burners, no such direct heat exchange from the flame or hot gases is possible. Examples of these burners are radiant tube furnaces, immersion tubes, flameless (dark radiant) gas burners and radiant elements.

Direct-fired or open flames burners may be further classified as non-premixed (*non-aerated*), partially premixed (*partially aerated*) and premixed (*aerated*)

© The Author(s), under exclusive license to Springer Nature Switzerland AG 2022
V. Raghavan, *Combustion Technology*,
https://doi.org/10.1007/978-3-030-74621-6_4

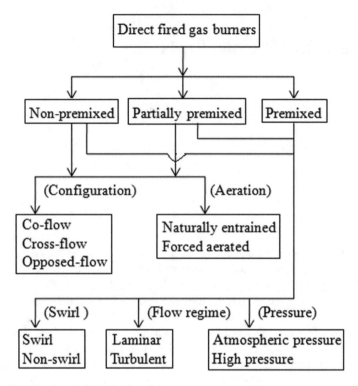

Fig. 4.1 Classification of direct-fired gas burners

burners. Further, classifications are made based on the operating pressure of the fuel gas, such as *atmospheric* or *high-pressure* burners, based on the method for air supply, such as *entrained air* burner or *forced aerated or ventilated* burners, and whether the flames are *laminar* or *turbulent*. Classifications of direct-fired burners are shown in a flowchart in Fig. 4.1.

In non-aerated burners, fuel and oxidizer/air flow into the combustion chamber through separate ports. Based on the directions of fuel and air flow, these burners are classified as *co-flow* burners, where the air flows parallel to the fuel stream, *cross-flow* burners, where air flow is perpendicular to the fuel stream and so on. Mixing of fuel and air is typically the rate-limiting step in these combustion systems. The mixing process and the flame extents depend on the flow rates of fuel and oxidizer, number of slots used to supply fuel and oxidizer, orientation of the streams and the presence of turbulence inside the combustion chamber. Burners involving jet diffusion flames are also used in selected applications, where longer flames are required, such as in cement kilns.

In partially aerated burners, a given amount of air, called *primary* air, equal to some percentage of the stoichiometric air requirement, is also fed along with the fuel stream. In general, this fuel–air mixture in the fuel stream will not be flammable, and an additional air is required for sustaining the combustion reaction. The additional air is termed as the *secondary* air and is supplied through separate ports. The configurations chosen may be same as those in non-aerated burners. Partially aerated burners have several advantages; they are cleaner, have sufficient operating range, generally higher than aerated burners, and have higher controllability. Therefore, these are used in many applications including domestic appliances and industrial furnaces.

Fully aerated burners employ flammable mixtures with the required equivalence ratio. These are used in special applications where a wide operating range is not required, and it is sufficient to establish stable flames within a narrow range of operating parameters. For example, due to the need of high temperatures in welding and metal cutting applications, acetylene–oxygen flammable mixture is employed in welding torches, where a high-temperature premixed flame is produced.

Based on the supply pressure of the fuel gas, burners may be classified as atmospheric- or high-pressure burners. Atmospheric-pressure burners operate at a gas supply pressure (gauge) in the range of 500–3000 Pa. Practically all domestic and commercial gas appliances fall in this category. On the other hand, high-pressure burners are used in combustion chambers having large volumes and in those where high heating rates are required. Some examples are industrial furnaces and kilns. In general, the flames resulting from low flow rate atmospheric pressure burners are laminar and those with large flow rates used in high-pressure applications are turbulent in nature.

In general, any burner design attempts to effectively mix the fuel and air in the required proportion at the fastest possible rate, in order to sustain the flame continuously for uninterrupted operation. However, depending upon the intended application, burners also differ in the manner in which these objectives are accomplished. These seemingly contradictory traits are realized in practice by starting from one design out of a limited set of simple, basic designs and then refining it to suit the desired application. A co-flow burner, a swirl burner and an entrained air burner operated at atmospheric pressure are the basic burners selected for discussion in the following sections.

A *co-flow* burner, where the fuel and oxidizer flow parallel to each other, may be operated in non-premixed, partially premixed and even in premixed modes. A simple burner of this type has a core port, typically a duct through which the fuel gas is fed, and an annular or co-flow port surrounding the core port, through which the oxidizer is supplied. A typical co-flow burner may be inverted to have oxidizer flow from the core port and fuel flow from co-flow port, which imparts an entirely different structure to the flame. In another configuration, a partially premixed fuel–air mixture may be fed through the core port and the secondary air through the co-flow port to impart increased control on the burning characteristics. In yet another configuration, if the core port is supplied with a rich reactant mixture and the co-flow port is supplied with a lean reactant mixture, resulting in the formation of a triple flame, which offers increased stability. Co-flow burners with multiple fuel and oxidizer ports are also available.

In a *swirl* burner, the supply of fuel or oxidizer or both is along a circumferential direction with respect to the cross-section of the burner. This enables faster mixing of reactants and also provides controllability to adjust the flame extents in both the axial and radial directions. The thermal efficiency of domestic appliances generally improves when the reactant flow is swirl-assisted.

In *atmospheric entrained air* burners, fuel gas is supplied into a mixing chamber through a nozzle or an orifice. Due to the momentum of the fuel jet, air from the atmosphere is naturally entrained into the mixing chamber. This partially premixed reactant flows through a burner head, which has multiple holes, in general. Secondary air also entrains naturally into the flame zone formed over the burner head. Such burners are extensively used in domestic and commercial cooking appliances.

4.2   Co-flow Burners

In order to improve the combustion characteristics of a simple jet diffusion flame, an air stream flowing through an annular area around the fuel jet is provided resulting in the so-called co-flow configuration. This enables an increased rate of transport of air to the flame over and above that due to the natural entrainment of quiescent air from the atmosphere. Therefore, the length of the resultant flame in the co-flow burner is shorter than that of the jet diffusion flame for the same fuel flow rate. Characteristics of flames from different types of co-flow burners are discussed next.

Consider the co-flow burner shown in Fig. 4.2. The entrainment of atmospheric air is completely suppressed in this burner owing to the presence of the confining wall on the outer periphery of the oxidizer stream. For a given fuel flow rate, if the oxidizer flow rate is high enough, the resultant flame has a shape similar to that of a jet diffusion flame and is called an *over-ventilated* flame. If the oxidizer flow rate is less, then an *under-ventilated* flame, where the flame stretches toward the oxidizer stream, is seen. These flames were first analyzed by Burke and Schumann. From an under-ventilated flame, as the co-flow rate of oxidizer is increased, the flame

Fig. 4.2 Schematics of flames from co-flow burner; fuel is supplied through core pipe and oxidizer through the annular region

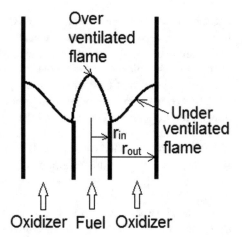

transforms to an over-ventilated flame and with a further increase in the oxidizer flow rate the flame extents are seen to decrease.

Numerical simulations of a laboratory-scale co-flow burner for different operating conditions have been carried out in order to gain insights into the characteristics of flames established in such burners. Methane is supplied in the core port, and air is supplied in the outer annular port. A detailed chemical kinetics mechanism involving 18 major and minor species has been used in order to accurately model the chemical kinetics of the combustion process. All the thermo-physical properties are calculated as functions of temperature and species concentrations. In order to account for flame radiation effects, an algebraic radiation model, which utilizes the optically thin approximation, is used. The internal diameter of the core port is 4 mm, and the internal diameter of co-flow pipe is 30 mm. This co-flow pipe is sufficiently long enough to create a confined flame. The flow rate of methane is maintained at 2.8×10^{-6} kg/s for all the cases. The flow rate of co-flow air is set to 4.5×10^{-6} kg/s to obtain an under-ventilated flame and increased to 63.6×10^{-6} kg/s to sustain an over-ventilated flame. Figure 4.3 presents the grayscale temperature contours and the contour corresponding to $\phi = 1$ (thick black line) in order to delineate the flame zone. The average velocities of fuel and air streams are shown in Fig. 4.3.

The case of a jet diffusion flame in the absence of the co-flow duct has been simulated for the same fuel rate. In this case, natural entrainment of air from the ambient takes place. One more case of over-ventilated flame with higher air flow rate of 127.2×10^{-6} kg/s has also been simulated. Figure 4.4 shows the grayscale temperature contours and contour line of unity equivalence ratio for jet diffusion flame along with two over-ventilated co-flow flames. It is apparent that the length of the jet diffusion flame is higher than those of over-ventilated diffusion flames for the same fuel flow rate. The radial extent of the jet diffusion flame is also higher than the over-ventilated flames. The co-flow air effectively decreases the flame volume. For the two over-ventilated flames in Fig. 4.4, there is a clear decrease in the radial extent of the flame and a slight increase in the maximum flame temperature due to an increase in the co-flow air supply. The flame length, however, remains the same for these two cases.

The effect of using pure oxygen in the co-flow instead of air is shown in Fig. 4.5 for a methane–oxygen co-flame established in a laboratory scale burner. First, the co-flow burner is considered to be unconfined, such that the exits of the fuel jet and the annular jet are in the same plane and open to the atmosphere. This is in contrast to the ones considered previously, in which the co-flow pipe was long enough to avoid any entrainment of ambient air. The internal diameter of the central fuel pipe is 2 mm, and that of the outer pipe is 16 mm. The methane flow in the central pipe is set to 4.5 L per hour (LPH). Pure oxygen is used in the co-flow pipe. When the oxygen flow rate in the annular jet is increased from 4.5 LPH to 9 LPH, and further to 13.5 LPH, there is an apparent decrease in flame length as seen in Fig. 4.5.

However, the visible flame extent in the radial direction has not changed notably. Further, the co-flowing oxygen imparts additional brightness to the edges of the over-ventilated diffusion flames, which is one of the primary features seen when pure oxygen is used as oxidizer. The brightness of the edges also increases with the

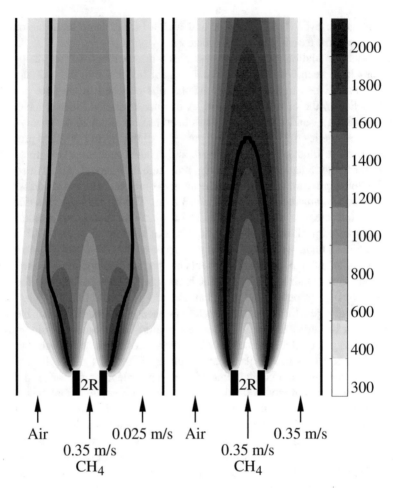

Fig. 4.3 Numerically predicted isotherms and contours of unity equivalence ratio of under- and over-ventilated flames

increased availability of the oxygen. Numerically predicted isotherms for these cases are presented in Fig. 4.6. Here, the numbers on the contour lines indicate the levels. Number 1 corresponds to minimum temperature and 8 corresponds to the maximum temperature. The values of the minimum and maximum temperatures are indicated in square brackets below each contour plot.

It is clear that as the oxygen flow rate increases from 4.5 LPH to 9 LPH, the maximum temperature increases from 2265 to 2637 K, due to the increased availability of pure oxygen and the occurrence of oxygen-enhanced combustion. Also, the maximum temperature region elongates and becomes an inverted "U"-shaped region as against a ring-shaped localized region seen for the lower oxygen flow rate of 4.5 LPH.

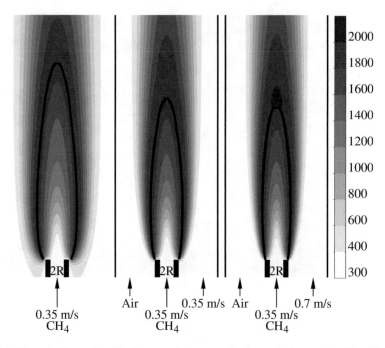

Fig. 4.4 Numerically predicted isotherms and contours of unity equivalence ratio of jet diffusion flame and over-ventilated flames

Fig. 4.5 Flame photographs of over-ventilated methane–oxygen co-flow diffusion flames; methane flow from central pipe is 4.5 LPH; and oxygen flow rate in annular region is varied from 4.5 LPH to 13.5 LPH

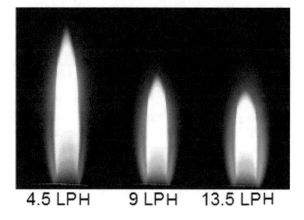

With a further increase in oxygen flow rate to 13.5 LPH, the maximum temperature increases to 2894 K and the maximum temperature region remains an inverted "U"-shaped region. However, the axial extent of the maximum temperature region is reduced indicating the occurrence of a rapid oxy-fuel reaction very near the fuel port. As the exiting fuel experiences the enhanced heat transfer from the flame, the minimum temperature region around the burner exit becomes smaller in size with

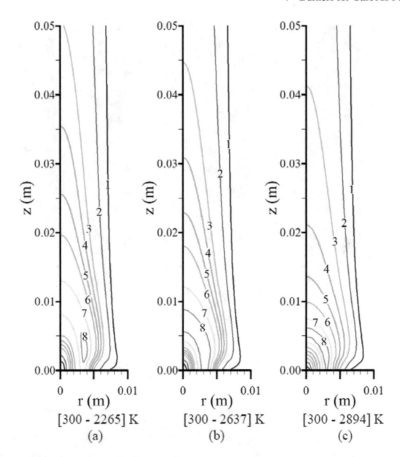

Fig. 4.6 Predicted isotherms in open co-flow oxygen–methane diffusion flames. Fuel flow rate is 4.5 LPH, and oxygen flow rates in LPH are **a** 4.5, **b** 9 and **c** 13.5

increasing oxygen flow rate. The use of improved radiation models and detailed chemical kinetics with a large number of species allows the maximum temperature value to be predicted quite accurately. In Figs. 4.3, 4.4 and 4.6, only a relative comparison between different cases is provided. It has also not been possible to measure the flame temperatures since these are quite high.

The effect of confinement on the features of co-flow flames, especially when oxygen is used in the co-flow is illustrated next. The same oxygen-methane co-flow burner has been enclosed at the bottom of a cylindrical confinement chamber made out of glass and having a length of 300 mm, which is 150 times the core pipe diameter. Its internal diameter has been varied from 44 to 176 mm. When the methane and oxygen flow rates are kept the same at 4.5 LPH, as the confinement diameter is increased from 44 to 176 mm, the flame length is seen to increase from 34 to 41 mm, as shown in Fig. 4.7.

Fig. 4.7 Effect of confinement diameter on flame height; fuel flow rate is kept as 4.5 LPH and oxidizer flow rates of 4.5 LPH and 13.5 LPH are considered

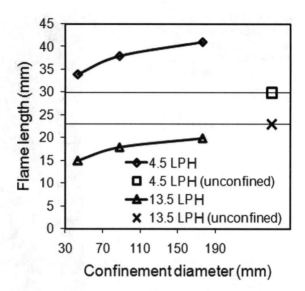

For these flow rates, the unconfined flame length is around 30 mm, which is observed to be the smallest. On the other hand, keeping fuel flow rate the same, when the oxygen flow rate is increased to 13.5 LPH, the flame heights obtained are 15 mm, 18 mm and 20 mm, respectively, for the confinement diameters of 44 mm, 88 mm and 176 mm. These are smaller than the unconfined flame length, which is around 23 mm. The variation in the flow field due to the variation in the diameter of the confinement forms the reason for the above trends.

Primary gaseous fuels obtained naturally, such as natural gas (methane) and LPG, are extensively used in many applications. Till now, characteristics of flames fuelled by methane have been discussed. Methane is lighter than air and, in general, burns in a clean manner. Methane produces soot only under certain oxygen-starved conditions. On the other hand, LPG is heavier than air. In general, it produces soot when burnt in a non-premixed or a partially premixed mode, due to its multi-fuel composition. Proper aeration technique is required to avoid soot formation in LPG flames. This is best demonstrated by an experimental study of LPG diffusion flames established from a typical co-flow burner. The core pipe has an internal diameter of 10 mm and the co-flow pipe internal diameter of 42 mm. The burner is kept open to the atmosphere. LPG is supplied at a rate of 6.25×10^{-6} kg/s through the core pipe. The air flow rate in the co-flow annular region has been varied from 100 to 200% of the stoichiometric air flow rate ($= 9.68 \times 10^{-5}$ kg/s) required for the set fuel flow rate.

Figure 4.8 presents instantaneous photographs of LPG–air diffusion flames. It is apparent that highly luminous flames result from the combustion of LPG in non-premixed mode. It is also clear that the flame luminosity in this case is quite different from the brightness of methane–oxygen co-flow flames presented in Fig. 4.4. Luminosity in LPG flames is predominantly due to soot radiation. Also, smoke at the tip of the flames shows that the soot particles have not been completely oxidized within

Fig. 4.8 LPG–air co-flow
diffusion flames with air flow
rates in the co-flow pipe
varying from 100 to 200% of
the stoichiometric air flow
rate required for the given
fuel flow rate

100%   150%   200%

the flame zone and are escaping. Further, even when the air flow rate is increased to
around 200% of the stoichiometric value, both flame length and flame radius do not
exhibit any noticeable changes, especially near the base of the flames.

In order to control the combustion characteristics in such sooty flames, some
amount of primary air may be added along with the fuel. This has been done in the
aforementioned experimental setup, while keeping the LPG flow rate the same as
in the previous cases and the annular air flow rate set at 200% of the stoichiometric
requirement. Figure 4.9 shows the flame photographs for different cases with varying
flow rates of the primary air indicated as a percentage of the stoichiometric air requirement. The mixture coming out of the core pipe is not flammable when the primary air
content is less than 40% of the stoichiometric value. Hence, the combustion process
in such cases occurs in the non-premixed mode. It is evident from Fig. 4.9 that the
flame height decreases significantly with the addition of primary air. The case with
40% primary air almost represents the limit of formation of premixed flame. The
equivalence ratio of the core mixture with primary air content at 40% of the stoichiometric value is around 2.5, calculated by assuming LPG to contain 60% butane and
40% propane by mass. As the primary air content increases to 50% (equivalence ratio
2) and beyond, the mixture ignites without requiring any secondary air. Once rich
premixed combustion starts, the flame height does not change significantly. Laminar
flame speed, rather than the flame height, becomes important for these cases.

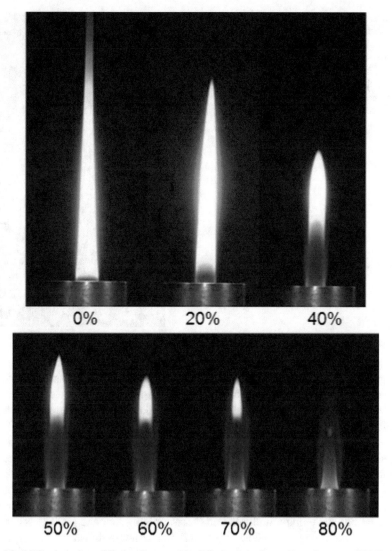

Fig. 4.9 LPG–air co-flow diffusion flames with air flow rate in the co-flow pipe set as 200% of the stoichiometric air flow rate required for the given fuel flow rate; primary air flow rate is varied as a percentage of the stoichiometric air requirement

As the primary aeration is increased, the extent of the bluish zone within the flame increases accompanied by the formation of an inner premixed flame cone. For 80% primary air, the equivalence ratio is around 1.25, and a clear inner premixed flame cone is visible along with an outer diffusion flame.

The co-flow burner may also be used to burn two streams of premixed fuel–air mixtures. For example, if a fuel-rich mixture and a fuel-lean mixture are supplied through the core and co-flow ducts, respectively, a triple flame is formed. Figure 4.10

Fig. 4.10 Methane–air triple
flame formed over the
co-flow burner. Rich
methane–air mixture ($\phi =$
2.4) and lean mixture ($\phi =$
0.4) supplied through core
and co-flow pipes,
respectively

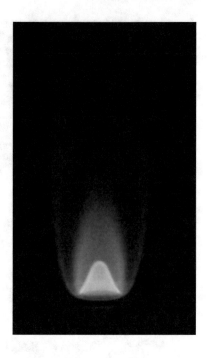

presents a photograph of a triple flame obtained by supplying a rich methane–air
mixture with an equivalence ratio of 2.4 in the core pipe and a lean methane–air
mixture with an equivalence ratio of 0.4 in the co-flow pipe. It is clear that a rich
inner premixed flame, surrounded by a diffusion flame and a lean premixed flame
enveloping both are clearly visible in Fig. 4.10.

From the above illustrations, it is clear that a co-flow burner offers increased
controllability for non-premixed flames and can be used to establish partially
premixed and premixed flames, including triple flames. All the flames presented
so far are laminar. By increasing the burner size or gas flow rates or both, turbulent
flames having varying characteristics may be sustained in co-flow burners. In general,
industrial burners employ multiple ports for supplying fuel and air. The schematic
of a typical multiple port co-flow turbulent flame burner is shown in Fig. 4.11.

Multiple port burners offer a wider operating range as well as flexibility. Better
control is achieved by changing the fuel and air supply rates from multiple ports,
thereby increasing the mixing and the stability of the combustion process. Primary
air may also be added to the fuel stream in order to obtain partially premixed flames.
In the air stream, pure oxygen may be added to initiate oxygen-enhanced combustion. The flame temperature may be controlled by introducing inert species such
as nitrogen. Co-flow burners are employed in many applications owing to these
advantages.

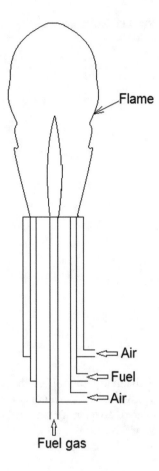

Fig. 4.11 Schematic of multiple port co-flow turbulent flame burner

4.3 Swirl Burners

Flame stability is essentially controlled by the degree of mixing of the fuel and oxidizer and the time of residence of the reactant mixture within the combustion chamber. In turbulent flame burners, where the fuel and oxidizer are supplied at high speeds, flame stabilization is achieved by strategic placement of bluff bodies in the flow field, by introducing pilot flames and by creating swirl. Stabilization of reaction zones using swirl is often employed in industrial burners and gas turbine combustion chambers. Using swirl flow, stabilization can be achieved in typically all the modes of combustion, namely non-premixed, partially premixed or premixed. In premixed combustion, swirl enhances the mixing of reactants and products, and in partially premixed and non-premixed modes, in addition it enhances the mixing of the fuel and oxidizer.

Figure 4.12 presents the schematic of a simple swirl burner, where a secondary air stream is injected circumferentially into the fuel and primary air streams. By

Fig. 4.12 Schematic of a
swirl burner

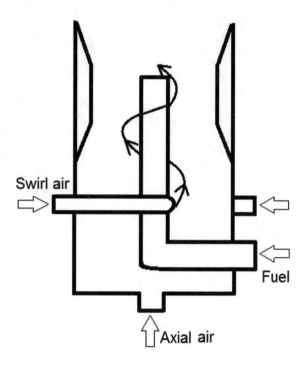

individually controlling the flow rates of primary and secondary air streams, the
strength of the swirl component can be enhanced and flame characteristics may be
altered. To quantify the swirl component, swirl number, S, as defined by Chiger and
Beer, is used. Here,

$$S = \frac{G_\theta}{RG_z} = \frac{\int_0^R wur^2 dr}{R \int_0^R uur\, dr}$$

Here G_θ and G_z stand for axial fluxes of azimuthal and axial momentum, respectively, u and w are the axial and azimuthal velocity components, and R is the outer
radius of the annular pipe. Swirl number may also be defined in simple terms as
the ratio of the momentum of the azimuthal velocity component to the momentum
of the axial velocity component. The basic effect of swirl component is to create
recirculation zones. A value of swirl number less than 0.4 is termed *low swirl*, and a
value higher than 0.6 is termed *high swirl*. Swirl component may also be introduced
using guide vanes and twisted tapes. In case a swirl or tangential vane is used, swirl
number may be defined using the geometrical parameters also.

Figure 4.13 shows the schematic of a typical flow field in a swirl burner. It may
be noted that there are multiple recirculation zones induced by air flow and fuel flow.
Based on the strength of the swirl component, the flame radius increases and the flame

Fig. 4.13 Typical flow field
in a swirl burner

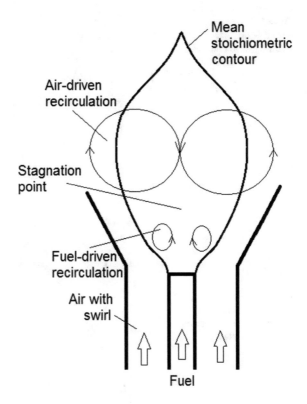

Mean
stoichiometric
contour

Air-driven
recirculation

Stagnation
point

Fuel-driven
recirculation

Air with
swirl

Fuel

height decreases, as indicated by the schematic contour of the mean stoichiometric ratio.

Schematics of typical flame shapes for different swirl numbers are presented in Fig. 4.14. It is apparent that the flame is quite long when there is no swirl component (Fig. 4.14a). In this case, the fuel has to travel a longer distance before the sufficient amount of air required for combustion is available. Furthermore, in this case, the mixing of fuel and air is mainly controlled by radial diffusion. As swirl is introduced, the flame radius increases and the flame length decreases (Fig. 4.14b). It may be noted that the internal vortices as shown in Fig. 4.13 may burst and cause elongation of the flame in axial direction as shown in Fig. 4.14b in flows with intermediate swirl. This type of flame is a combined swirl and jet flame. Mixing of the fuel and oxidizer as well as reactants and products is enhanced with the addition of the swirl component. With high swirl, flames with closed internal circulation, as shown in Fig. 4.14c, are obtained. Typically in industrial burners, a swirl number of 0.5 is customarily employed to control the flame extents.

The typical flow field in a swirl burner employing premixed reactants is shown in Fig. 4.15. The recirculation zone present inside the flame zone enhances the mixing of products and reactants. This helps in the reduction of pollutants such as NO_x, apart from increasing the stable operation regime of premixed flames. Combustion of lean

Fig. 4.14 Schematics of
typical flame shapes for **a** no
swirl, **b** low to intermediate
swirl and **c** high swirl

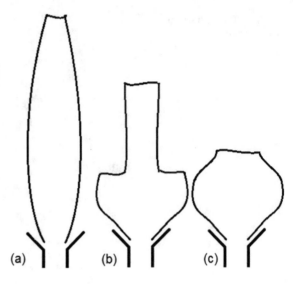

Fig. 4.15 Flow field in a
swirl burner with
circumferentially injected
premixed reactants

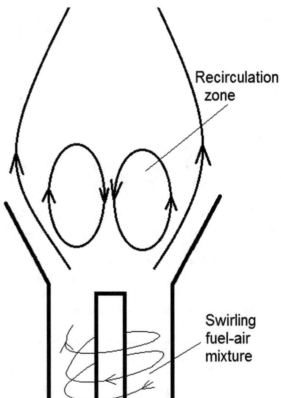

fuel–air mixture is also possible in such burners. In high-speed burners, a pilot flame is usually established in a small annular region at the burner exit to sustain the main flame. The flow rate involved in the pilot flame is typically around 5% of the flow rate in the main flame.

In industrial burners, adjustable swirl vanes are employed to control the swirl number, and therefore, the flame extents. These vanes are fitted to the annulus portion of a co-flow burner. Fuel with primary air is fed through the core pipe, and secondary air is fed through the annular portion through the swirl vane.

4.4 Atmospheric Entrained Air Burners

An atmospheric entrained air burner operates at atmospheric pressure on well-known Bunsen burner principle. Fuel gas is fed through an orifice or nozzle into a mixing tube as shown in Fig. 4.16. Due to the momentum of the fuel jet, air from the atmosphere is entrained through the adjustable holes provided. Fuel and air mixes in the mixing tube. The mixture thus formed is usually rich in fuel. The rich reactant mixture leaves through the ports of an annular burner head. The diameter of the port and the separation distance between any two ports are fixed such that, when ignited, almost individual flames are formed in the ports. Secondary air is entrained from the central hole of the annular burner head as well as from the exterior of the burner top portion, as shown in Fig. 4.16.

Typical pressure distribution in the burner is shown in Fig. 4.17. Gas from a gas-line or high-pressure cylinder is regulated and sent into the injector at a pressure higher than the atmospheric pressure. The pressure rapidly decreases as the fuel is accelerated through the orifice. The pressure attains a local minimum at the throat region. As a result of this pressure drop, atmospheric air entrains into the mixing tube due to the Bernoulli effect through the holes provided. In the mixing tube, the

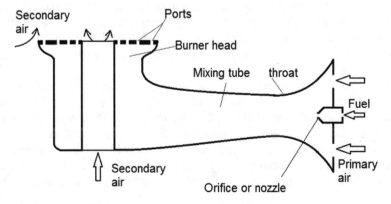

Fig. 4.16 Schematic of an atmospheric partially aerated burner

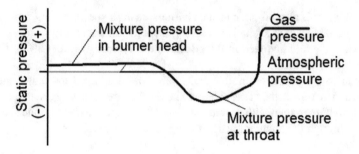

Fig. 4.17   Typical pressure distribution in atmospheric burner

pressure recovers and attains a value slightly higher than the atmospheric pressure. The mixture pressure in the burner head is also slightly higher than atmospheric pressure, so as to push the reactant mixture through the ports. The flame pressure is very close to the atmospheric pressure.

Turndown ratio is the ratio of the maximum to minimum fuel flow rates achievable in a given burner. Referring to Fig. 3.14, the maximum fuel flow rate is limited by the degree of completion of the combustion process, blow-off, and by the amount of primary air that can be entrained. The minimum fuel flow rate is limited by the flashback limit of the burner. In general, for most gas burners of this type, a turndown ratio of 5 to 1 is desirable.

4.5   Stability, Performance and Emission Characteristics

As already discussed, burners employing premixed flammable reactants have instability problems such as lift-off, flashback and blow-off. These instabilities adversely affect the performance of the burner resulting in poor combustion and reduced operating range. In partially premixed burners, any perturbation, namely an increase or decrease of the supply pressure of gas during operation, alters the flow rate of the reactant mixture, possibly leading to unstable operation. In general, flame extinction, lift-off or blow-off may occur under such conditions leading to poor performance and excessive formation of pollutants.

In co-flow and swirl burners, depending upon the velocity of the core (fuel) jet or the co-flow (oxidizer) jet, flame may lift-off. Partial premixing, enhancing the oxygen content in the oxidizer and addition of highly reactive fuels such as hydrogen are strategies employed to expand the stable operation regime as well as enhance the performance of these burners. In confined chambers, when the temperature is high enough and excess air is also available, the formation of nitric oxides is inevitable. This may be mitigated to some extent by providing partial recirculation of products around the flame zone. However, this must be done with care, as too much recirculation of the products may deplete the oxygen availability for the fuel, and this may result in the formation of unburned hydrocarbons, soot and CO.

In general, in forced aerated burners, since the supply of both fuel and air are metered, it is possible to achieve better control over the performance and pollutant formation. On the other hand, in burners where air naturally entrained solely due to the momentum of the fuel jet, likelihood of unstable operation is higher and only a limited operating range is usually achievable. Therefore, atmospheric entrained air burners are adjusted to operate at certain gas pressures, within a narrow range, decided based on the fuel used. Testing is carried out to confirm that the burners are capable of stable operation in the pressure range prescribed, without requiring any adjustment during the operation. The tests should ensure steady operation without any flame lift-off or formation of yellow flame tips. Flame lift-off is due to increase in the air entrainment and yellow tipping is mainly due to insufficient primary air entrainment. The tests should further ensure that only negligible (well within the standards) amount of CO leaves along with the burnt gases. The orifice size or the fuel manifold pressure is adjusted so that the fuel flow rate conforms to the rated heat input of the burner. The size of the holes for primary air inlet may be adjusted by using a shutter. By doing so, a proper proportion of primary air is allowed to entrain and required flame characteristics in stable operation regime are obtained. Based on the input power to the burner (fuel flow rate) and the quantity of the entrained primary air, different regimes, similar to those shown in Fig. 3.14, are obtained. This is shown in Fig. 4.18. Even though the dimensional values of the quantities plotted may differ from one burner to another, the relative shapes of the curves remain the same for all the burners employing different types of fuels. Entrainment of excessive amount of primary air will result in the operating regime being above line AB in Fig. 4.18, and the flame then tends to lift-off from the burner ports. Under certain circumstances, the flame may blow-off. Primary air should be controlled using the shutter under such conditions. On the other hand, if the entrainment of primary air is not sufficient enough, then the operation regime shifts below the line CD, resulting

Fig. 4.18 Operation regimes of a burner based on input power and primary air

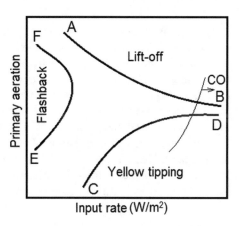

in long yellowish flames from the burner ports. This is indicative of incomplete combustion accompanied by the formation of CO and soot. Increasing the opening of the shutter to entrain more primary air rectifies this issue. If, for some reason, the supply pressure of the fuel gas, and hence, the flow rate decreases, the operating regime may fall to the left of line EF. In such case, the flame may either extinguish or may propagate into the burner port into the mixing tube to find more reactants.

Flashback results in the formation of soot, heating of the orifice and burner head, and the extinction of the flame accompanied by noise. The flashback limit indicated by line EF depends on several factors such as type of fuel, number of ports, port size and temperature of the burner head. For example, coke-oven gas has a bigger flashback zone than that of natural gas. Cooling of the burner head is usually done to shrink the extent of the flashback zone. From the above discussion, it is clear that, for stable operation, the operating regime should be well within the bounding lines AB, CD, EF and the line that indicates the emission of CO along with hot gases.

4.6 Design Procedure

In forced aerated burners, the momentum fluxes of the fuel and air flow streams entering the combustion or mixing chambers are kept almost the same. Since the mass flow rate of fuel is much less than the mass flow rate of air, fuel is injected with a higher velocity through an orifice or a nozzle. For selecting an orifice, the ratio of its length to its diameter and the angle of approach form the important parameters. Several standard orifices are available. The coefficient of discharge, C_d, of an orifice is determined experimentally. The value of C_d varies from 0.8 to 0.95 for an orifice injector typically used in burners. If p is the pressure of the gas upstream of the orifice (in N/m^2), the volumetric flow rate of the fuel (in m^3/s) is given as,

$$\dot{V} = C_d A_o \sqrt{\frac{2p}{\rho_F}}$$

Here, ρ_F is the density of the fuel (in kg/m^3) and A_o is the area of opening of the orifice (in m^2). The mass flow rate of the fuel (in kg/s) is determined by multiplying the volumetric flow rate by fuel density. The input heat supply rate into the burner (in kW) is determined by multiplying the mass flow rate of the fuel by its calorific value (in kJ/kg). In general, the lower calorific value of the fuel is used. Further, keeping in mind some combustion efficiency in the range of 0.8–0.9, the mass flow rate of the fuel is increased to achieve the overall design power input to the burner.

In forced aerated burners, the required mass flow rate of air is supplied through another port either into a mixing chamber or directly into the combustion chamber, based on the mode of combustion desired. The total air supply is split to meet the primary air and secondary air requirements, based on the design. In practice, some excess air is supplied, usually in the secondary air stream, to ensure complete combustion. With these metered flow rates of fuel and air, desired flame zones may be established in the combustion chamber built using proper materials, with proper insulation and flame holding devices.

The amount of air naturally taken inside an entrained air burner, as shown in Fig. 4.16, depends on the fuel orifice area (A_o), throat area (A_t) and the total burner port area (A_p). The area of opening for air entrainment is usually kept adjustable and will be in the range of 1.8 to 2.5 times the port area. The ratio of volumetric flow rate of air that has entrained to the volumetric flow rate of the fuel is termed as *entrainment ratio*, E. If r is the *relative density* (density of the fuel gas divided by the density of the air), then using Prigg's formula, for same air and fuel inlet temperatures, an ideal value of E may be calculated as,

$$E = \sqrt{r}\left[\sqrt{\frac{A_p}{A_o}} - 1\right].$$

If the fuel and air temperatures are different, then E may be calculated using a modified expression taking into account the temperature of air (T_a) and fuel (T_F) streams entering the mixing chamber, as shown:

$$E = \sqrt{r}\left[\sqrt{\frac{A_p T_a}{A_o T_F}} - 1\right].$$

In an actual scenario, there will be losses in orifice, burner ports and diffuser. The coefficient of discharge of flame ports, C_{dp}, which is in the range of 0.6 to 0.7, is lower than the C_d of the fuel orifice, which is usually in the range of 0.85–0.95. Due to the presence of diffuser in the mixing tube (Fig. 4.19), a frictional loss component, called the friction loss coefficient, C_L, should also be considered. The length of the diffuser (L_m) is proportional to the throat diameter (d_t) and is usually 12 times the diameter of the throat. The value of C_L depends upon the length as well as the included taper angle of the diffuser. Its value varies between approximately 0.14, when the included

Fig. 4.19 Mixing tube in atmospheric entrained air burner

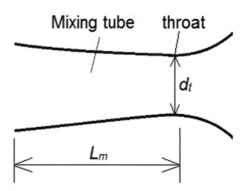

angle is between 5° and 8°, to approximately 0.45, when the angle is 20°. If the angle is reduced below 5°, say to 3°, C_L increases to around 0.18. Therefore, in general, the taper angle is fixed in between 5° to 8°. Using C_{dp} and C_L, the actual entrainment ratio may be calculated as,

$$E = \sqrt{\frac{r C_{dp}}{1 + C_L} \left(\frac{A_p}{A_o}\right)} - \frac{1 + r}{2}.$$

For an ideal case, C_{dp} is equal to unity and C_L is equal to zero.

In general, for the design of atmospheric entrained air burners, area of the throat, A_t, is taken as 50 to 100 times the fuel orifice area, A_o. Similarly, the burner port area, A_p, is taken as 60 to 150 times A_o. Usual design practice is to consider the ratio of A_t to A_p in the range from 0.65 to 0.8. Using these ratios and experimentally determined range of C_{dp} and C_L, the entrainment ratio may be calculated using,

$$E = \sqrt{\frac{r A_p}{A_o} C_1} - \frac{1 + r}{2}.$$

Here, C_1 is an empirical constant that varies between 0.5 and 0.65. The above calculations are used to determine the fuel and air flow rates into the burner. The geometrical parameters are set such that the value of E is in the range of 4–6 for having 40–60% primary aeration.

Design aspects of the burner port are discussed next. In forced aerated burners, depending upon the speed of flow of the gases, the burner head is engineered with ceramic tiles having diverging area, other flame holding devices such as bluff bodies and pilot flame ports. Two such designs are schematically shown in Figs. 4.20 and 4.21. To help control blow-off when the gas flow speed is high, a flame holder, or bypass, as shown in Fig. 4.20, may be employed. A flame holder operates on the general principle that a flame can be maintained in a zone of relatively low velocities, where the residence time for the reactants to react is higher. Further, if a steady flame is maintained in the zone of low velocity of the flame holder, it would cause rapid

Fig. 4.20 Schematic of a flame holder

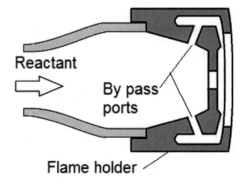

Fig. 4.21 Schematic of a burner with a burner tile and tangential pilot flame port

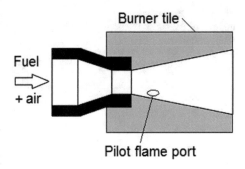

ignition of the fresh reactant mixture coming in contact with it and exiting through the main part of the burner.

As already mentioned, hot ceramic surfaces are helpful in sustaining the flame. Such ceramic surfaces, which can withstand high temperatures and become incandescent, are called burner tiles. In order to burn the fast-flowing reactants, often a pilot flame, established using lean fuel–air mixture fired tangentially over the surface of the burner tile, is used. Such a burner is schematically shown in Fig. 4.21.

In a fully aerated (premixed) burners, flame arrestors are provided at appropriate locations. An operation regime based on the design turndown ratio is fixed, and a stable operation is ensured. Usually, the flame exits through a single port in these types of burners. On the other hand, in entrained air burners and domestic stoves, multiple flame ports are usually provided. In multi-port burners, at low aeration, typically up to around 65% of primary aeration, flames from neighboring ports interact and a single inner cone may be established over an array of ports. As the primary aeration is increased to around 75%, individual inner flame cone above each port may become visible. These are shown in Fig. 4.22.

In the range of primary aeration between 65 and 75%, the single cone becomes oscillatory or even unstable and when the primary aeration is more than 75%, the cones oscillate and may even become unstable, depending upon the fuel flow rate. Therefore, for multi-port burners, the regimes may be schematically represented as in Fig. 4.23.

Fig. 4.22 Illustration of combined and individual flames in multi-port burners

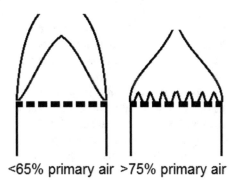

Fig. 4.23 Schematic of
regimes of flames in
multi-port burner

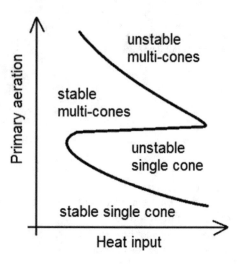

Port loading is the heat input to each port, expressed in W/mm². The flame port
should be designed based on the operation regime determined using port loading as
one of the parameters. Unless an external stabilization mechanism is used, experimental investigations suggest that typically for burners operating with 50% primary
aeration, the port loading should be in the range of 9–14 W/mm². For burner head
having ports of circular cross-section, an increase in the diameter of the port results
in increased lift-off stability; that is, the tendency to lift-off is reduced. For a burner
head having several smaller-sized ports, the value of C_{dp} is lower. Thus, with the same
total burner port area and heat input, the burner port loading increases for smaller
ports, which may cause the flame to lift-off. Therefore, several smaller ports may be
replaced by a fewer larger sized ports. The port diameter, on the other hand, affects
the flame length directly. Therefore, based on the requirements of flame length, port
loading, and the primary aeration range to be employed in the burner, a proper port
diameter may be chosen. For rectangular ports with a given slot width, an increase
in the port length decreases the tendency of the flame to lift-off. The thickness of
the burner plate is generally fixed less than 2 times the port diameter. Ports that are
deeper than this may experience fluid dynamics induced instabilities.

Inter-port spacing is another important parameter in a multi-port burner. The inter-
port spacing should be such that there is a favorable interaction between the flames
in terms of enhanced heat transfer and oxygen availability for centrally located or
inner rows of ports. Oxygen supply is the reason for the increase in the overall flame
length from multiple ports when compared to that of the single port as the inter-
port spacing is reduced. This issue may be overcome by staggering the ports. On
the other hand, when the inter-port spacing increases, the heat interaction decreases
and this increases the tendency of the flame to lift-off. Therefore, both inter-port
spacing and positioning of the ports must be selected in such a way as to ensure
steady operation. In general, spacing (between the nearest edges of the ports) more

than 6 mm is considered high and spacing in the range from 1.5 mm to 6 mm is seen to approximately double the primary aeration at which the flame lifts-off.

The salient points of the design procedure for gas burners that are discussed above are summarized below for the sake of convenience:

1. The maximum heat input for the selected application is calculated based on the calorific value of the fuel used.
2. The fuel gas injection pressure and dimensions for a suitable standard orifice required to provide the calculated heat input to the burner are estimated.
3. The air flow rate required for the burner, which usually includes the primary and the secondary air, is calculated based on the type of fuel used. In the case of forced aerated burners, proper inlet port area for primary air supply should be provided to mix the fuel and air efficiently. In the case of swirl burners, depending upon the desired swirl number, the azimuthal component of fuel/air stream is supplied appropriately. In the case of naturally aerated burners, proper dimensions for the mixing tube and burner port are estimated.
4. A proper flame holding device must be selected to sustain the flame in high-speed applications. Pilot flames, ceramic tiles and by-pass ports are also used to hold the flames. In the case of multi-port burners, proper dimensions for the individual ports and inter-port spacing are selected based on the port loading.
5. In the case of fully aerated burners and in those where the likelihood of a flashback is high, flame arrestors at proper locations must be provided.

4.7 Several Design Concepts

In this section, design concepts of a few forced aerated burners, which novel design strategies to mix the fuel and oxidizer, are presented. Figure 4.24 presents the schematics of six burners. The fuel and oxidizer streams are represented by F and O, respectively. The fuel stream may have primary air or diluents based on the application. Figure 4.24a, c–d utilizes nozzle mixing concept. The fuel and oxidizer are mixed at the exit of the burner in these cases. Various parameters such as recess depth (Fig. 4.24a), angle of swirl vanes (Fig. 4.24c) and angle of the oxidizer ports (Fig. 4.24d) are used to control the mixing and flame stabilization processes. The recess depth may be adjusted by moving the fuel injector forward or backward, so that the degree of mixing within the burner exit may be varied. The angled ports direct the oxidizer flow toward the fuel stream to enhance the mixing process. The angle of the swirl vanes control the swirl number and this in turn controls the mixing of the fuel and oxidizer as well as the flame extents.

The designs shown in Fig. 4.24b, e–f utilize the internal mixing concept. Here, the fuel and oxidizer are mixed internally. Flame stabilization is accomplished using swirl vanes (Fig. 4.24b), angled ports oriented such that the reactant mixture takes a diverging path (Fig. 4.24e) and by providing tangential (circumferential) entries to both fuel and oxidizer (Fig. 4.24f). In addition, pilot flames may be introduced near the exit of these burners. An annular reactant stream for providing a steady pilot

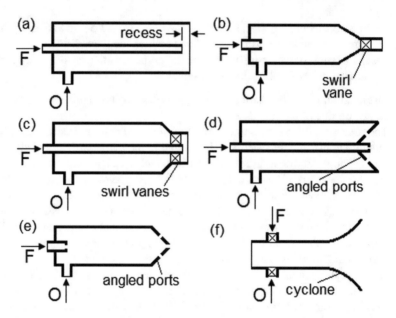

Fig. 4.24 Schematics of burners having different mixing schemes

flame can be used in the burners shown in Fig. 4.24b, e. Small pilot flames in the cyclone region may be employed in the burner shown in Fig. 4.24f.

From the above discussions, it is clear that many different types of gas burners are employed in industrial and domestic applications. Efficient mixing of fuel and oxidizer and flame holding in the desired operation regime are the important requirements in the burners. Industries use several types of fuels and interchanging the fuels within a given burner is an important challenge. Turndown ratio and stability are affected when a fuel type is changed in the burners. Therefore, at every stage, proper design calculations and experimentation are required to efficiently use the burners and to have increased controllability and minimum emission of pollutants. There are a few burners that use both oil and gas fuels. Some features of such burners are discussed in the next chapter.

Review Questions

1. List the parameters based on which gas burners are classified.
2. Define over-ventilated and under-ventilated flames.
3. Explain the effect of increasing co-flow air supply in over-ventilated flames.
4. Comment about the changes occurring in the reaction zones when oxygen flow rate is increased.
5. How swirl number is calculated when swirl vanes or twisted tapes are used?

Exercise Problems

1. Assuming LPG as 60% butane and 40% propane by volume, calculate the primary air flow rate at which the mixture will become flammable. If hydrogen is mixed with LPG in volumetric fractions of 25, 50 and 75%, calculate the primary air requirement to meet the rich flammability limits in these mixtures.

2. Design an LPG (60% butane and 40% propane by volume) fueled burner for a cooking stove that has a number of circular ports arranged in a circle. The burner must deliver 2.4 kW at full load and operate with 40% primary aeration. For stable operation, the loading of an individual port should not exceed 12 W per mm^2 of port area. Also, the full-load flame height should not exceed 25 mm. Determine the number and the diameter of the ports.

3. Design an atmospheric entrained air burner using synthetic gas as the fuel. Composition of synthetic gas may be taken as 13% CO, 8% H_2, 15% CO_2, 3% CH_4 and rest N_2.

4. If biogas (65% methane and 35% CO_2 by volume) has to be used in an atmospheric entrained burner designed for LPG with the same power rating, list the changes to be made (if any) in the orifice, diffuser throat and burner head, relative to the existing values.

5. In a co-flow swirl burner, propane with 20% primary air is sent through the central core pipe at a rate so as to achieve a power rating of 5 kW. The annular region for the co-flow is fed with air at a rate so as to achieve an overall equivalence ratio of 0.4 inside the furnace. The co-flow air goes through the twisted tape that imparts 50% tangential component, 10% radial component and 40% axial component. Calculate the mass flow rates of fuel, primary air, secondary air and swirl number.

Chapter 5
Burners for Liquid Fuels

As mentioned in earlier chapters, liquid fuels are either vaporized or atomized into droplets before they are injected into a combustion chamber. The vaporization and atomization characteristics of the liquid fuel, the required rate of burning including the turndown ratio that fixes the operation range and the possible dimensions of the combustion chamber are key aspects to be considered in the design of a liquid fuel burner. The important properties associated with liquid fuels were discussed in Chap. 1. The vaporization and burning characteristics of a liquid fuel droplet were discussed in Chap. 3. This chapter starts with a discussion of different types of liquid burners used for the combustion of highly volatile fuels that do not require atomization. A discussion of atomization process and spray combustion aspects are presented next. Design and operation methodologies are discussed toward the end of the chapter.

5.1 Types of Liquid Fuel Burners

In this section, different types of liquid fuel burners and stoves are presented. The terms burner and stove may be used interchangeably, since size is the only difference between them. However, it is customary to use the term burner in the context of industrial applications and the term stove in the context of domestic applications.

Liquid fuel burners may be classified based on the mechanisms employed to vaporize the liquid fuel and to mix the fuel vapor with oxidizer, as follows:

Wick burners
Pre-vaporizing burner
Vaporizing burner
Porous burner
Atomizing burner

© The Author(s), under exclusive license to Springer Nature Switzerland AG 2022
V. Raghavan, *Combustion Technology*,
https://doi.org/10.1007/978-3-030-74621-6_5

5.1.1 Wick Burner

Wick burners are used as lamps and cooking stoves in rural areas and camping stations. Here, a wick is used to supply the liquid fuel in a container to the flame zone. Generally, the wick is made up of a fabric material such as cotton and it is also consumed, however, at a much slower rate. Typically, liquid fuels such as kerosene, alcohol and vegetable oils are used in wick burners. The working principle involved in a wick burner is quite similar to that of a candle. A thin fuel film is formed on the surface of the wick that is saturated with the fuel and is heated by the flame initiated at the tip of the wick. The liquid film evaporates and a diffusion flame sustains at the tip of the wick. Liquid fuel flows up through the wick by capillary action and feeds the fuel to the flame zone. In a wick burner, the wick projects out of a guide tube or a sleeve. The length of the wick projecting out is controlled by a wick-lifting mechanism. At the exit of the guide tube, a diffusion flame is anchored over the vaporizing film of liquid fuel formed on the surface of the wick. Also, a perforated metallic enclosure is kept surrounding the flame zone up to a particular height. This is heated up by the flame and also helps in preheating the entraining air. As a result, the burning performance is improved. Alternatively, a flat wick of rectangular cross section is also employed in place of a cylindrical wick. A wick burner is shown schematically in Fig. 5.1.

In a wick stove, usually fuels such as kerosene and alcohols are employed. Here, multiple cylindrical wicks are mounted in a circular arrangement each within its own guide tube. Two coaxial, perforated, thin cylindrical tubes, called inner and outer sleeves, surround the wicks with a small clearance between the edge of the wick and the sleeves. The wick assembly can be momentarily raised from its initial position, through the annular space in between the sleeves, by using a lever. By doing so, the

Fig. 5.1 Schematic of a wick burner

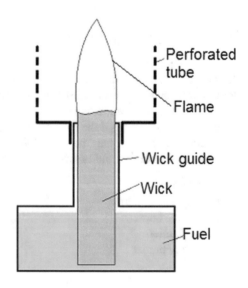

wicks, saturated with the liquid fuel, can be ignited using a pilot flame. After ignition, the lever is released so that the wicks reach their initial position. The fuel vapors, generated due to heat transfer from the flame to the wicks, flow upward, entrain the atmospheric air through the perforations in both the inner and outer sleeves and burn in a partially premixed mode. The flame subsequently heats up the sleeves and the wicks and enables continuous vaporization of the liquid fuel from the surfaces of the projected wicks. A fuel-rich flame exists within the annular portion between the sleeves. However, a proper bluish flame is established at the top of the burner due to continuous entrainment of air through the sleeves into the fuel-rich partially premixed flame. An adiabatic cylindrical outer wall encloses the outer sleeve with an adequate gap. The height of this outer wall is more than that of the sleeves. The efficiency of the wick stove can be improved by reducing the heat losses. A wick stove is shown schematically in Fig. 5.2. In an improved device, the perforations and the annular gap is designed such that the fuel–air mixture burns only at the top of the sleeves after being mixed with secondary air and no flame exists within the annular space between the sleeves. The heat is transferred to the liquid fuel at the surface of the wick by the hot sleeves.

Some of the disadvantages of the wick burner/stove are that the burning rate is slow and only light and volatile fuels can be used since the capillary action may not be strong enough in the case of heavy oils to draw the fuel up to the tip of the wick for sustaining the combustion.

Fig. 5.2 Schematic of a wick stove

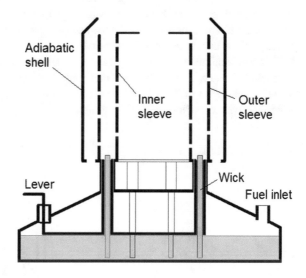

5.1.2   Pre-vaporizing Burner

In this type of burner, volatile liquid fuels such as methanol, ethanol, gasoline, kerosene and naphtha are pre-vaporized in separate chambers or pipelines and the resultant vapors are supplied to the combustion chamber. A small heat exchanger is used to vaporize the liquid fuel. Liquid fuels are allowed to pass through tube or coil, which is heated by the hot combustion products prior to leaving the combustion chamber. The heat exchanger is designed such that the liquid fuel vaporizes at a steady rate at which the vapor is supplied to the combustion chamber. This is accomplished by providing the necessary sensible heat and the latent heat of vaporization. Pre-vaporizing burners are employed in applications such as blow torch, pressurized liquid fuel stove, gas turbine combustion chambers and lean premixing pre-vaporizing combustors.

A sketch of a simple pre-vaporizing burner is shown in Fig. 5.3 in order to illustrate its working principle. Here, a liquid fuel from a pressurized tank is fed through a coiled metallic tube, which passes through the flame zone. In a domestic application, the liquid fuel tank may be pressurized using a small hand pump. A primer fuel such as an alcohol or naphtha is used to create a diffusion flame, which heats up the fuel carrying tube and the nozzle chamber initially. Fuel vaporizes and exits through the nozzle with a high velocity. The momentum of the vapor jet naturally entrains air from the atmosphere and the fuel vapor and air mix thoroughly. A dispersion plate is kept at the top of the burner to deflect the mixture radially. Flames are initiated above the dispersion plate and heat up the vaporizer tube passing over them. As a result, increased amount of fuel vapor exits the nozzle, entraining even more air. After some time, a steady burning regime is established.

The key challenge in this design is vaporizing the liquid fuel completely before it reaches the nozzle and ensuring that fuel vapor alone comes out of the nozzle. The characteristics of the flame is controlled by adjusting the opening area of the air holes and by controlling the fuel feed rate.

Fig. 5.3   Sketch of a simple pre-vaporizing burner

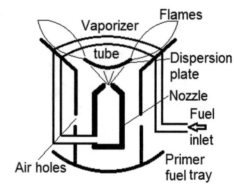

Fig. 5.4 Pot vaporizing
burner

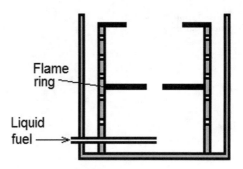

Flame
ring

Liquid
fuel

5.1.3 Vaporizing Burner [2]

In this type of burners, a liquid fuel is directly fed into the combustion chamber, where it vaporizes and subsequently burns. Vaporizing burners are used in air heaters, water and room heating boilers. Kerosene and alcohols are the common fuels used in these burners. It is essential that the liquid fuel be vaporized very quickly soon after entering the combustion chamber. To facilitate this, a thin film of liquid fuel is formed over the hot surfaces within the combustion chamber. A film of liquid fuel may be formed by using a pot, a centrifugal spinner or a rotating cup and a wick. With this definition, a wick burner may also be classified under the category of a vaporizing burner. Typically asbestos or ceramic wicking agents are used in vaporizing burners to produce a fuel film. Sufficient amount of oxidizer is provided into the combustion chamber by using natural or forced draught mechanisms.

An example of a vaporizing burner is a pot-type burner. A pot burner comprises of an oil reservoir and combustion chamber as shown in Fig. 5.4, slowly introduced into the pot along the surface of the pot. As the liquid flows down under the action of gravity, it forms a thin film along the surface of the pot. This film is evaporated by the radiant energy received from the combustion chamber walls and the flame. The fuel vapors rise in the pot and mix with primary air naturally entraining through the holes in the sides of the inner walls. Under normal operating conditions, a flame sustains just above the flame ring. These burners are less efficient and may also be sooty under certain conditions. Forced aeration is used to improve the performance and emission characteristics. Carbon deposits are also inevitable in pot burners. Therefore, provision for cleaning should be available and periodic cleaning of the burner is essential. These burners are used in applications where the power requirement ranges between 15 and 25 kW.

5.1.4 Porous Burner [2]

In Chaps. 1 and 3, flameless premixed combustion in a porous inert medium was briefly mentioned. The so-called radiant burners, which are fully aerated inert porous

Fig. 5.5 Schematic of a
porous burner

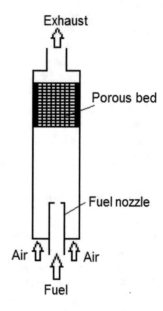

burners and utilize variety of gaseous fuels, achieve high combustion efficiency and
low emissions. When a liquid fuel is employed in such burners, enhanced vaporization
followed by high intensity combustion (high radiant output) is achieved. The porosity
provides both increased surface area as well as a longer residence time for evaporation
and burning processes.

A basic porous burner is schematically shown in Fig. 5.5. Initially, the bed of
porous material is heated up using a volatile primer fuel. The liquid fuel is sprayed
into the bottom of the hot porous bed, so that it penetrates into the pores and vaporizes
within the bed.

The required air supply also reaches the porous material as shown in Fig. 5.5. A
flame zone is formed inside the porous bed and is sustained by a continuous supply
of fuel and air. The distance between the spray nozzle and the porous material is one
of the important control parameters, other than the material and porosity of the bed,
spray characteristics and the burner rating. Usually ceramics such as magnesium
stabilized zirconium, silicon carbide and yttrium-stabilized zirconium are selected
as porous materials, and they are manufactured to have a porosity of around 80–85
percent. In general, the CO and nitric oxide emissions from this type of burner are
low.

5.1.5 Atomizing Burner

The burners described so far may be used only with moderately or highly volatile
liquid fuels. It is difficult to vaporize heavy oils such as diesel and fuel oils of

different grades at a steady rate at which they are intended to burn. Also, these oils are multi-component in nature and display non-uniform evaporation characteristics. The practice is therefore to break up the liquid fuel into small droplets, such that the overall surface-to-volume ratio increases thereby resulting in faster evaporation. For example, if a droplet of diameter of 10 mm is broken into 10^6 droplets each of diameter 100 μm (0.1 mm), the surface area increases by a factor of 100. Atomization is the process used for disintegrating a liquid into small droplets, which can evaporate and burn at the required rate in a spray combustion chamber.

Atomization technique can be used with different types of liquid fuels. As mentioned in Chap. 1, the main requirement to successfully atomize a liquid fuel is that the liquid viscosity should be within a permitted value. Atomizers are the devices used for generating droplets having certain size range and number density. The resultant suspension of fine droplets in a surrounding gas is termed as a spray. Atomization is achieved in different ways. A liquid fuel is forced through an opening of small area such as an orifice under high pressure. This results in the disintegration or the breakup of the liquid jet coming out of the orifice into small droplets. In another method, a two-phase mixture of the liquid fuel and a gas or steam is injected at high velocities. The interaction between the two fluids results in the disintegration of the liquid into tiny droplets. In all the methods, the breakup of the liquid essentially occurs as a result of instability associated with the motion of the liquid jets relative to the surrounding gas or vapor. The sizes of droplets formed and their distribution are functions of the type of liquid fuel, size and configuration of the atomizer, nature of flow, gas and liquid velocities and pressure. The droplet sizes and the size distribution obtained from an atomizer are generally determined from cold injection studies. The important properties of a spray of droplets are the spray angle, droplet number and size distributions and droplet velocities.

In a spray, the fuel vapor released from the evaporating droplets mixes with the oxidizer and burns either as a diffusion flame or a premixed flame, depending upon whether a flame is present around the droplet or not. When a droplet is quite small, say less than 25 μm in diameter, experiments reveal that no flame surrounds the droplet and the droplet rapidly vaporizes. On the other hand, if the droplet is bigger, an individual flame may surround the droplet and heterogeneous combustion takes place. In the case of volatile fuels, where the atomization usually results in small droplets, premixed combustion is seen. However, there are notable differences between the propagation of a premixed flame through a homogeneous gaseous fuel–air mixture and a spray with small rapidly vaporizing droplets, primarily because of the non-uniform distribution of the droplet diameter. The basic features associated with liquid fuel atomization are discussed next.

5.2 Basics of Atomization [10, 19]

Atomization of a liquid into discrete droplets is accomplished by methods involving aerodynamic forces, mechanical forces, ultrasonic forces and so on, either alone or as a combination. Cohesive forces in a liquid stream must be overcome in order to disintegrate a liquid jet into small droplets. This may be accomplished in several ways. Relative motion between the liquid jet and the surrounding gas can cause instabilities on the surface of the jet, which grow and eventually break up the jet. Surface tension at the interface between the liquid and the gas plays a key role in this breakup process. Centrifugal forces that arise within a swirling liquid jet can help in disintegrating the liquid jet. Externally applied mechanical, electrostatic and acoustical forces can also cause disintegration of the liquid jet. Atomization process requires a certain energy to initiate an instability in a liquid stream. The growth of the instability causes a breakup of the liquid jet eventually leading to the formation of droplets.

When a liquid jet issues from a nozzle, it interacts with the ambient fluid and oscillations or perturbations occur on the surface of the jet. The velocity gradient across the liquid–gas (ambient) interface causes the perturbations to grow as this is a Kelvin–Helmholtz instability. While viscosity damps the perturbation, surface tension can cause amplification or attenuation of the perturbation, depending on the wavelength. Eventually, the liquid jet breaks up into droplets. The continuous length of a liquid jet prior to break up and the droplet size are important parameters in a liquid jet breakup process.

According to Weber's theory, in circular jets, it is assumed that any disturbance causes rotationally symmetrical oscillations on the jet surface. Surface forces may either damp out or amplify the disturbance, depending upon the wavelength of the oscillation. If the disturbance is amplified, the jet may eventually break up. For steady injection of a liquid into a quiescent gas through a single injector with a circular orifice, the mechanisms of jet breakup are typically classified into four primary regimes depending upon the relative importance of inertial, surface tension, viscous, and aerodynamic forces. Each regime is characterized by the magnitudes of the Reynolds number, Re_L, and Weber number We_L, defined based on the properties of the liquid. A dimensionless number called Ohnesorge number (Oh or Z), defined as a function of Reynolds number and Weber number, is generally used to characterize the regimes. A Weber number (We_A) defined based on the density of the ambient gas (ρ_A) is also used in characterizing the regimes. These non-dimensional numbers are defined as follows:

$$Re_L = \frac{\rho_L U_L d_o}{\mu_L}; \ We_L = \frac{\rho_L U_L^2 d_o}{\sigma}; \ We_A = \frac{\rho_A U_L^2 d_o}{\sigma}; \ Oh = Z = We_L^{0.5} Re_L^{-1}.$$

Here, U_L is the jet exit velocity, d_o is the exit diameter of the orifice, μ_L is the liquid viscosity and σ is the surface tension. Table 5.1 lists the atomization regimes, predominant breakup mechanism and values for the non-dimensional numbers in each regime. An illustration of the breakup of a liquid jet in each of these regimes is given in Fig. 5.6.

When a liquid emerges from an orifice very slowly, droplets of almost uniform size are formed at a regular rate due to the surface tension forces (Fig. 5.6a). In wind-induced breakup regimes, oscillations of the liquid jet lead to the formation of

Table 5.1 Regimes of atomization

Regime	Breakup mechanism	Criteria
Rayleigh jet breakup (Varicose breakup)	Surface tension force	$We_A < 0.4$ or $We_A < 1.2 + 3.41\ Z^{0.9}$
First wind induced breakup (sinuous wave breakup)	Surface tension force, dynamic pressure of ambient gas	$1.2 + 3.41\ Z^{0.9} < We_A < 13$
Second wind induced breakup (wave like breakup with gas friction)	Dynamic pressure of ambient gas	$13 < We_A < 40.3$
Straight atomization (formation of droplets right from the jet exit)	Possible mechanisms are aerodynamic interaction, turbulence, cavitation, bursting and so on	$We_A > 40.3$ or $Z \geq 100(Re_L)^{-0.92}$

Fig. 5.6 Regimes of atomization: **a** Rayleigh jet breakup, **b** first wind-induced breakup, **c** second wind-induced breakup and **d** straight atomization

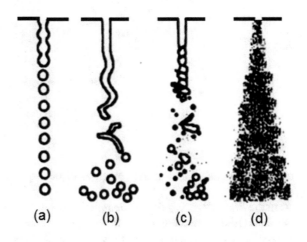

(a)   (b)   (c)   (d)

ligaments, which further break into droplets (Fig. 5.6b, c). These droplets interact with each other farther downstream and result in a secondary atomization. At much higher jet velocities, formation of droplets begins right from the jet exit (Fig. 5.6d). In the case of non-circular injectors (slot injectors), a sheet of liquid rather than a jet issues out of the injector. This sheet is unstable even to small perturbations and is stretched by the interaction with the ambient fluid to form films. These films roll up into cylindrical ligaments and breaks up into droplets.

5.2.1 Types of Atomizers [19]

The most commonly used methods for atomizing liquid fuels are (a) *pressure* atomization, (b) *rotary* atomization and (c) *twin-fluid* atomization. These are shown pictorially in Fig. 5.7. These methods may be generally classified into two major categories

Fig. 5.7 Schematic
representation of commonly
used atomizers: **a** pressure
atomizer, **b** rotary atomizer
and **c** twin-fluid atomizer

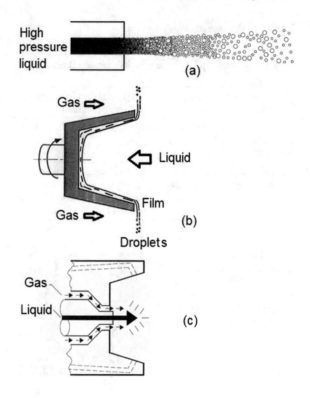

based on the magnitude of the velocities of the liquid stream and the surrounding gas.

In the first category, a liquid stream with a high velocity emerges into a quiescent or relatively slow-moving gas, as in pressure atomization and rotary atomization. In the second category, a relatively slow-moving liquid comes into contact of a stream of gas flowing at high velocity as in twin-fluid atomization. Application-specific atomization techniques which do not fall into these categories are: (a) effervescent atomization, (b) electrostatic atomization, (c) ultrasonic atomization and so on. The salient features commonly used atomizers are discussed next.

5.2.1.1 Pressure Atomizers

In these atomizers, high pressure is used to force the liquid fuel through a nozzle. The liquid emerges in the form of a high velocity jet, generally into a quiescent gas environment. Due to the instabilities on the surface of the jet, it disintegrates into droplets. Atomizers that utilize this basic idea but differ in the actual design are:

1. Plain orifice
2. Simplex
3. Duplex

4. Dual orifice
5. Spill return
6. Fan spray

Plain orifice atomizer is the simplest of all the pressure atomizers in terms of fabrication and operation. However, its spray quality is poor at low-to-moderate injection pressures, and high injection pressures are required in order to obtain a spray with an acceptable quality.

A rotational velocity component or swirl may be imparted to the liquid jet in order to distribute the droplets over a larger spray volume. This is implemented in the *simplex* type of pressure atomizers, also called *pressure-swirl* atomizers. The liquid fuel enters through conical or helical or circumferential slots and has a tangential or azimuthal component of velocity. These configurations are shown in Fig. 5.8. The liquid fuel is discharged through an orifice as an annular film into the ambient. This film breaks into droplets due to friction between the surface of the film and the ambient gas. A wider spray angle is obtained.

Improvements in simplex atomizers are required for obtaining a better spray quality. *Duplex*, *dual orifice* and *spill return* atomizers, which are improvised versions of simplex atomizers, are schematically shown in Fig. 5.9. Duplex type of pressure

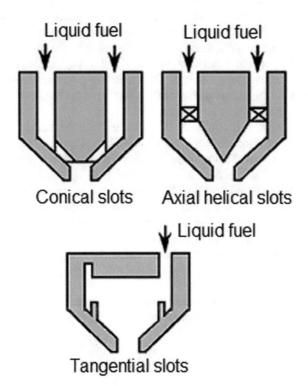

Fig. 5.8 Schematics of simplex atomizers

Fig. 5.9 Schematics of duplex, dual orifice and spill-return atomizers

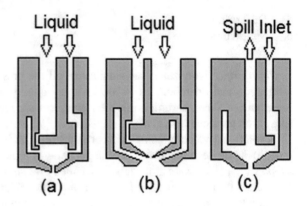

atomizers is similar to simplex type, but with two inlets for feeding the liquid fuel, called the primary and secondary inlets. The supply pressure in both the inlets is maintained at the designed value. The flow rate through the primary inlet is smaller than that through the secondary inlet. The dual liquid fuel streams interact within the injector and come out of the orifice as a single stream.

Dual orifice type of pressure atomizer consists of two simplex type of injectors fitted concentrically one inside the other. The injector which is mounted inside is called the primary orifice. The two injectors are placed such that the spray from the primary orifice interacts with the spray from the secondary orifice just outside the atomizer. When the required fuel flow rate is low, fuel is supplied through the primary injector alone. Fuel is injected through the secondary injector, when the fuel flow rate requirement is high. The quality of the spray from the dual orifice atomizer is good when both the injectors are in operation.

Spill return atomizer is similar to the simplex atomizer, except that there is a passage at the rear to allow certain quantity of liquid fuel to return from the atomizer without being atomized. This passage is called the spill-return passage. The inlet valve to the atomizer is operated at the maximum design pressure, which allows for the maximum design liquid flow rate. However, a valve in the return line is operated in such a way to admit a given quantity of the liquid through the return line and only the required mass flow rate of liquid is injected. Due to the high pressure operation, the spray quality is much better in this atomizer.

Fan spray atomizers are used for spray coating. In these, the spray is formed by impinging a round liquid jet on a curved surface, resulting in a narrow elliptical spray pattern.

A comparison of the performance, applications, advantages and limitations of different types of pressure atomizers is provided in Table 5.2.

Table 5.2 Comparison of different types of pressure atomizers

Type	Droplet size in microns	Applications	Advantages	Limitations
Plain orifice	25–250	Diesel engines, jet engine after burners, ramjets	Simple, rugged, cheap	Narrow spray angle
Simplex	20–200	Gas turbines, industrial furnaces	Simple, cheap, wide spray angle	High supply pressure
Duplex	20–200	Gas turbine combustors	Simple, cheap, wide spray angle, good atomization	Narrow spray angle with increasing flow rate
Dual orifice	20–200	Variety of aircraft, gas turbines	Good atomization	Complexity in design
Spill return	20–200	Variety of combustors	Wide spray angle, good atomization	Higher power requirements
Fan spray	100–1000	High pressure paint coating	Good atomization	High supply pressure

5.2.1.2 Rotary Atomizers

In these atomizers, liquid fuel is fed on to a rotating surface. A uniform film of liquid spreads out as a result of the centrifugal force. A *flat disc* or a *cup* is generally used as the rotating surface. The surface is ensured to be smooth and vibration-free during operation. The disc may have vanes or slots. The rotational speeds are such that the centrifugal force is higher than gravitational force. These atomizers are larger in size than the pressure-swirl atomizers for the same liquid flow rate. Different processes occur on the surface of the rotating flat disc depending upon the liquid flow rate and the rotational speed. Direct droplet formation, formation of both droplets and ligaments, ligament formation subsequently accompanied by generation of droplets, film or sheet formation and its disintegration into small droplets are the typical regimes observed in rotary atomization. The atomization quality is improved by increasing the rotational speed, decreasing the liquid feed rate, heating the liquid to reduce its viscosity and creating serrations on the outer edge of the cup or the disc. A comparison of the performance, applications, advantages and limitations of different types of rotary atomizers is provided in Table 5.3.

5.2.1.3 Twin Fluid Atomizers

In twin fluid atomizers, a high velocity gas stream impinges on a slow flowing liquid sheet or jet. In general, air or steam at high velocities is often used. These injectors are also called *air-assist* or *steam-assist* atomizers. The liquid fuel encounters the gas or steam either inside or just outside the atomizer. The former is called an internal

Table 5.3 Comparison of different types of rotary atomizers

Type	Droplet size in microns	Applications	Advantages	Limitations
Flat disc	10–200	Spray drying, aerial dispersion of pesticides, chemical processing	Monodisperse droplets, good control of spray quality and liquid flow rate	360° spray pattern
Cup	10–320	Spray drying, spray cooling	Can handle slurries	Possible requirement for air blast around the periphery

mixing twin fluid atomizer and the latter is called an external mixing twin fluid atomizer. In the case of air-assist atomizers, the air sent through the atomizer forms a part of the primary air required for combustion. For low viscosity fuels, external mixing twin fluid atomizers are employed. For high viscous liquid fuels, air or steam is mixed with the liquid inside the atomizer itself in order to improve the atomization process. Such internal mixing twin fluid atomizers are commonly used in furnaces. Power requirements of the two fluid systems are higher than that of the pressure atomizers. Various arrangements for external mixing and internal mixing atomizing burners are shown schematically in Figs. 5.10 and 5.11, respectively.

In the external mixing type, the liquid fuel is injected through one set of holes and is gusted by a high-velocity jet of steam or air issuing from the other holes. Mixing occurs entirely outside the atomizer. Typically a flat flame, with good combustion efficiency, is produced in moderate capacity burners. Fluctuations in the liquid fuel feed rate should be avoided in these burners. At high fuel feed rates, the fuel stream becomes highly dense. This hinders the good mixing of fuel and air in the furnace.

Fig. 5.10 Schematics of external mixing twin fluid atomizers

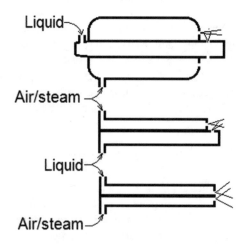

Fig. 5.11 Schematics of internal mixing twin fluid atomizer

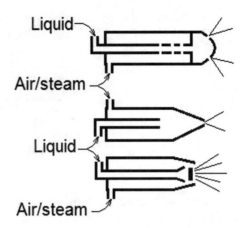

Liquid

Air/steam

Liquid

Air/steam

The internal mixing type of atomizer is more commonly used for heavy oils, as it provides high efficiency at high fuel feed rates. It also has a wide turndown ratio and a flexible flame shape. Depending upon the actual configuration, either a flat spray or a hollow cone spray is obtained. In most burners of this type, high-pressure air is generally used for atomization in place of steam, as air atomization results in more rapid, complete and efficient combustion. In case of steam-assist atomizers, about 0.7–5% of the total steam generated in the boiler is typically used in the burners for atomization at a pressure of 2.75–5.5 bar. The injection pressure of the liquid fuel is slightly higher than that of either the steam or the air. The liquid fuel need not be preheated as much for steam-assist burners as compared to the air-assist burners. The same procedure as that employed in pressure atomizing burners is used to supply the air required for combustion to the steam–liquid fuel mixture. Generally, a port of 25 mm diameter per 0.5 kg/h of fuel is used to feed air. It must be ensured that the air is not injected in a direction opposite to the spray penetration direction.

While an air-assist atomizer operates with small amounts of high velocity air, an *air-blast* atomizer uses fairly large quantities of air. The air thus supplied is used for both the primary and secondary air requirements in a combustion chamber. There are two types of air-blast atomizers; *plain-jet* type and *pre-filming* type. In the pre-filming type, the liquid is spread out in the form of a thin sheet, when it encounters the atomizing air blast. Figure 5.12 presents the schematic of a low pressure air-blast atomizing burner, which works with heavy grade fuel oil at low excess air and high turndown ratio. These burners are operated with air preheated to around 350 °C, using the heat of the exhaust gases, so that the overall thermal efficiency of the system is enhanced as a result of improved spray quality. These burners are commonly used in furnaces in glass and ceramic industries, billet heating, heat treatment, forging, galvanizing, ladle heating and tea dryers. The burner comprises of an inner pipe, through which the primary air is supplied, as shown in Fig. 5.12. This pipe has a swirl vane at its exit to impart a circumferential component of velocity to the primary air.

Fig. 5.12 Schematic of a low-pressure air-blast atomizing burner

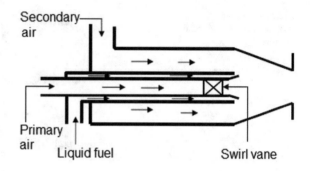

A very small annular gap is maintained between the outer surface of the inner pipe and the inner surface of another pipe mounted coaxially. The liquid fuel, generally a furnace oil, is fed through the uniform annular gap. The fuel flows out of the annulus as a thin film in the form of a hollow cylinder. Another larger concentric pipe is mounted above the two pipes in order to feed the secondary air as shown in Fig. 5.12. The hollow cylindrical film of the liquid fuel is thus sandwiched in between the streams of primary and secondary air. Due to the interaction between the liquid film and air, the liquid fuel film breaks into tiny droplets. A good degree of atomization is obtained even with highly viscous liquid fuels. This burner may be operated using gaseous fuels also and in a *dual fuel* mode, where both liquid and gaseous fuels are injected simultaneously. For dual fuel mode, the gaseous fuel is supplied through the primary air line. The dual fuel mode firing is utilized in applications, where there is not enough supply of gaseous fuels. In several installations, these burners handle liquid fuels at the rate of 100–200 kg/h and gaseous fuels at the rate of 10–400 m^3/h. Fuels such as light diesel oil, high-speed diesel oil, naphtha, high viscosity fuel oil, natural gas and LPG may be used in these burners. A turndown ratio of about 7:1 is achievable in these burners when high viscosity fuel oil is used.

A comparison of the performance, applications, advantages and limitations of different types of twin fluid atomizers are provided in Table 5.4.

Two atomizers, effervescent atomizer and ultrasonic atomizer, which are used in specific applications, are discussed subsequently. *Effervescent* atomizers are used in industrial oil burners. Here, gas or air bubbles are introduced into the liquid fuel. The bubbles rapidly grow when the bubbly liquid is discharged through the atomizer. The interaction between the bubbles and the liquid shatters the liquid fuel into fine droplets. Small droplet sizes and a favorable distribution, which can enhance mixing, are obtained from effervescent atomizers. These atomizers have large orifices and work without clogging while using heavy fuel oils.

Ultrasonic atomizers are commonly used in applications such as central heating facility. There are two types available. In the first type, the actuation of the atomizer surface using a piezoelectric device causes atomization. Such atomizers have a conical or flat vibrating surface, over which the liquid fuel is fed. In the second type, a sonic pressure wave is induced using a sonic nozzle. Here usually compressed air or steam passes through a convergent–divergent nozzle into a resonator chamber at

Table 5.4 Comparison of different types of twin fluid atomizers

Type	Droplet size	Applications	Advantages	Limitations
Air assist, internal mixing	50–500	Gas turbines, industrial furnaces	Good atomization, low risk of clogging, can handle high viscosity liquids	Requirement of large source of high pressure air
Air assist, external mixing	20–140	Gas turbines, industrial furnaces	Good atomization, low risk of clogging, can handle high viscosity liquids	High supply rate of external pressurized air
Air blast, plain jet	15–130	Gas turbine combustors	Simple, cheap, good atomization	Narrow spray angle
Air blast, pre-filming	25–140	Variety of aircraft, gas turbines	Good atomization, wide spray angle	Poor atomization at low velocities

supersonic velocities. The resultant high-frequency pressure wave is focused onto an open cavity. The energy in the acoustic wave atomizes the liquid fuel, which is continuously supplied to the cavity. High frequencies, varying between 6 and 100 kHz, are employed. A very high degree of atomization is achieved and droplet sizes varying between 1 and 25 μm are obtained. The main drawback is the requirement of a very high pressure.

5.2.2 Correlations for Droplet Diameters

The combustion of a liquid fuel spray occurs in three main steps: (1) droplet vaporization accompanied by heterogeneous combustion (diffusion flame around a liquid droplet) in some cases, (2) mixing of fuel vapors with the oxidizer and (3) homogeneous (premixed gas phase) combustion. The vaporization step is the slowest among the three steps. Heterogeneous combustion is faster than vaporization, but slower than the convection driven vapor-oxidizer mixing. The homogeneous combustion step is the fastest, as the reaction rates are quite high. In general, fine droplets (diameter lesser than 25 μm) can only vaporize and cannot be ignited to have individual diffusion flames around them. On the other hand, bigger droplets (diameter greater than 100 μm) vaporize due to the heat transfer from the surroundings and the fuel vapor is consumed in a diffusion flame established around the droplet. Therefore, in order to estimate the time required for vaporization and burning of droplets, both droplet sizes and distribution (number density) in a spray should be known quite accurately.

The size of droplets formed and the distribution of droplet sizes in a spray depend upon the type of the atomization process, the size of the atomizer and operating conditions such as pressure and feed rates of the liquid fuel and gas. In general, the droplet sizes and the size distribution are determined from flow experiments carried out in non-reactive environment. In the simplest possible experimental setup, molten wax is injected instead of the liquid fuel and the droplets formed in the spray are allowed to cool to form solid spheres. The droplet sizes and the size distribution are then analyzed. With the advent of laser-based diagnostics, optical visualization and related measurements are carried out along with high-speed imaging techniques to determine the droplet sizes and their distribution.

Empirical expressions for the droplet diameter distribution have been reported by researchers such as Tanasawa and Tesima, Nukiyama—Tanasawa and Rosin—Rammler. In its simplest form, the droplet size distribution may be written as,

$$\frac{dN}{dD} = a D^{\alpha} \exp\left(-bD^{\beta}\right), \tag{5.1}$$

where dN is the number of droplets having diameters between D and $D + dD$ and a, b, α and β are constants. It has been shown by Tanasawa and Tesima that in general α and β have the following values: $\alpha = -0.5$, $\beta = 1$ for a swirl atomizer; $\alpha = 2$, $\beta = 1$ for a pressure jet atomizer and $\alpha = 1$ and $\beta = 1$ for a twin-fluid atomizer. The parameters a and b are obtained from a curve fit of experimental data to Eq. (5.1).

Droplets involved in a spray have an assortment of sizes centered around a mean diameter. However, a theoretical analysis capable of predicting the mean diameter of the droplets produced by a given atomizer is not available yet and experimental data alone is used for this purpose. In a spray having N size ranges of droplets, if droplets of sizes $d_1, d_2, d_3, \ldots d_N$, are present, and the number of droplets in each of these sizes is $n_1, n_2, n_3, \ldots n_N$, respectively, a statistical mean value may be evaluated using the general formula given as,

$$d_{ab} = \left[\frac{\sum_1^N n_i d_i^a}{\sum_1^N n_i d_i^b} \right]^{1/(a-b)} \tag{5.2}$$

Based on the purpose or application, a set of values of a and b may be assigned and corresponding statistical mean values may be evaluated. Table 5.5 presents the definitions of different statistical mean values and the typical purposes for which they are employed.

Among these, the Sauter mean diameter, SMD or d_{32}, is the most widely used statistical mean, especially in combustion applications. Mathematically, it gives the mean diameter of a droplet, whose ratio of volume to surface area is the same as that of the entire collection of droplets. Research works have shown that for combustion applications, the SMD is the most appropriate metric for quantifying the fineness of a spray.

Empirical correlations are available for estimating the mean droplet diameters, as a function of the orifice size, supply pressure, liquid and gas properties. The physical quantities may be conveniently grouped in terms of non-dimensional quantities such as the Reynolds number and Weber number.

Table 5.5 Different methods to evaluate statistical mean value of droplets in a spray

Mean	Common name	a	b	Definition	Purpose
d_{10}	Arithmetic mean	1	0	$\left[\dfrac{\sum n_i d_i}{\sum n_i}\right]$	Estimated for relative comparison
d_{20}	Surface mean	2	0	$\left[\dfrac{\sum n_i d_i^2}{\sum n_i}\right]^{1/2}$	Estimation of mean surface area
d_{30}	Volume mean	3	0	$\left[\dfrac{\sum n_i d_i^3}{\sum n_i}\right]^{1/3}$	Estimation of mean volume
d_{21}	Surface area-based length mean	2	1	$\left[\dfrac{\sum n_i d_i^2}{\sum n_i d_i}\right]$	Used in applications involving absorption
d_{31}	Volume-based length mean	3	1	$\left[\dfrac{\sum n_i d_i^3}{\sum n_i d_i}\right]^{1/2}$	Used in applications involving evaporation and molecular diffusion
d_{32}	Sauter mean (SMD)	3	2	$\left[\dfrac{\sum n_i d_i^3}{\sum n_i d_i^2}\right]$	Used in applications involving mass transfer and combustion
d_{43}	Herdan mean	4	3	$\left[\dfrac{\sum n_i d_i^4}{\sum n_i d_i^3}\right]$	Used in applications involving combustion

Correlations for the mean droplet size (e.g., the SMD) for pressure atomizers are expressed in terms of the volumetric (or mass) flow rate of the liquid, $\dot{V}$, density of the ambient gas, ρ_g, and the pressure difference, Δp, in a general form as follows:

$$d_{32} = C_1 \rho_g^{n_1} \dot{V}^{n_2} \Delta p^{n_3},$$

where C_1, n_1, n_2 and n_3 are constants. Correlations for rotary atomizers depend on parameters such as the liquid feed rate, rotational speed of the cup or the disc, surface tension, density and dynamic viscosity of the liquid fuel. For twin fluid atomizers, correlations are expressed in terms of the orifice diameter, d_0, and the ratio of the mass flow rate of the liquid ($\dot{m}_L$) to the mass flow rate of the gas ($\dot{m}_g$). In a general, these are written as,

$$d_{32} = C d_0^m \left(\frac{\dot{m}_L}{\dot{m}_g}\right)^n.$$

Here, C, m and n are constants. The values of the constants in these correlations are usually obtained from a curve fit of the experimental data of the particular atomizer over a range of operating parameters. The droplet sizes and size distribution are important parameters used in the analysis of spray combustion. The basic aspects of spray combustion are discussed subsequently.

5.3 Introduction to Spray Combustion [9, 10, 19]

Spray combustion has been in practice right from the end of eighteenth century as a powerful method of burning non-volatile liquid fuels. In fact, spray combustion is the effective method of burning heavy fuel oils even today. The liquid fuel is present in the form of discrete droplets in a spray. The droplets have a range of sizes and they move in different directions with different velocities relative to that of the ambient gas. Since the droplets of varying sizes are spatially distributed non-uniformly, the resultant fuel vapor and oxidizer mixture are also spatially non-homogenous and non-uniform. This causes irregularities in the propagation of the flame through the spray. Spray flames observed in industrial furnaces are highly complex, as a result of the complex flow and mixing patterns in the combustion chamber, heat transfer processes involved and the non-uniform size distribution of the spray droplets. In order to understand the basics of spray combustion, idealized, dilute or coarse sprays with moderate sized droplets are considered.

The processes involved in the combustion of a dilute spray having moderate-to-large sized droplets (>100 μm) are shown in Fig. 5.13. These sprays have individual droplets burning with a diffusion flame surrounding them. The burning rate would vary depending upon the separation distance between any two droplets, droplet diameter at a given time instant and the relative velocity of the droplet with respect to the ambient gas.

As the droplets move through the spray, they vaporize as a result of heat transfer from the hot zone downstream. The fuel vapors surrounding the liquid droplet mix with the ambient oxidizer, usually atmospheric air. The reactant mixture around the droplet is heated up to the auto-ignition temperature, and a diffusion flame surrounds

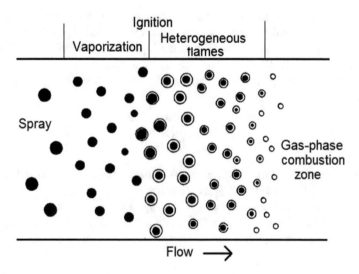

Fig. 5.13 Schematic of burning of coarse sprays with heterogeneous droplet flames

Fig. 5.14 Homogeneous combustion in dilute sprays with small droplets

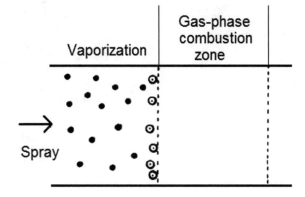

the droplet. The droplet burns almost steadily, with its diameter as well as relative velocity decreasing as a function of time. Once the entire droplet evaporates, the flame surrounding it extinguishes. A secondary homogeneous or gas-phase combustion zone is formed downstream.

On the other hand, if the droplet sizes in the spray are small (<25 μm), then the droplets approaching the flame zone vaporizes rapidly due to heat transfer from the flame and only the vapor mixed with ambient air reaches the combustion zone. In this case, there are no individual flames surrounding the droplets. Such an idealized configuration is schematically shown in Fig. 5.14.

Two idealized and extreme cases of dilute sprays have been discussed above. In an actual spray, based on the size of the combustion chamber and the type of the atomizer used, both small as well as large droplets are present with a given size distribution. Further, the spray may not always be dilute. Therefore, both heterogeneous and homogeneous reactions zones, distributed in a non-uniform manner within the combustion chamber, are present in such sprays. The physical processes involved in spray combustion are presented in Fig. 5.15.

When a liquid fuel is supplied at the required rate through an atomizer, droplets having certain range of sizes are formed based on the type of atomizer chosen. Smaller droplets vaporize rapidly as a result of heat transfer from the hot zones. The resultant fuel vapor mixes with the available air and homogeneous (gas phase) reaction zones are established. On the other hand, the bigger droplets undergo further collisions, where secondary atomization may take place. As the bigger droplets vaporize, the fuel vapor mixes with the ambient air, and upon ignition, diffusion flames surround the droplets. Gaseous products are formed in the homogeneous and heterogeneous reaction zones. As these products move out of the combustion chamber, the processes such as recombination, soot oxidation and equilibrium reactions take place. Additional processes also occur in the recombination zone when heavy oils are used. The droplets of heavy multi-component liquid fuels become highly viscous after they are heated and light components are released. They solidify into a porous coke. The diameter of the coke particle formed is usually around one-third of the initial diameter of the liquid droplet. The burning of a coke particle within the combustion

Fig. 5.15 Processes
occurring in spray
combustion

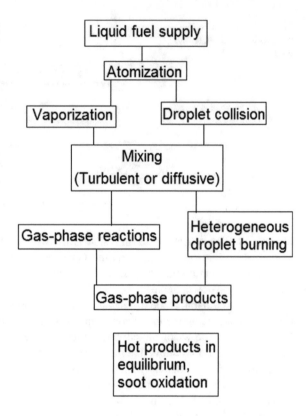

Fig. 5.15 Processes occurring in spray combustion

chamber is the most rate-limiting process in spray combustion of heavy oils. This is an important aspect and adequate residence time must be provided in order to allow the coke particle to burn out completely. Similarly, the vapors of heavier fuel fractions contribute to soot, since they have soot precursors. The design of the combustion chamber should facilitate the complete oxidation of coke and soot particles by having provisions for sufficient secondary air supply.

5.4 General Design Procedure

Every liquid fuel burner consists of an air supply system, a liquid fuel delivery system and an ignition system. The main chamber, where combustion occurs, must be designed very carefully. Slots and holes for air inlets, the passages for liquid and air flow have to be fabricated with the required precision so as to minimize if not avoid any blockage. The construction should be such that the burner assembly can easily be dismantled for carrying out maintenance operations in situ. Provisions for introducing swirl in the inlets, recirculation in the flow field and adjusting the spatial distribution of the spray are required to ensure complete combustion and to avoid particulate emissions. The combustion chamber dimensions should be such that an adequate residence time for the oxidation of CO and other hydrocarbons is provided. Also, depending upon the fuel and the oxidizer used, provisions for introducing additives for flue gas treatment such as limestone–water slurry and diluents to reduce sulfur dioxide and nitric oxide emissions should be made. The important points involved in the general design procedure of a liquid fuel combustor are given below:

1. The required flow rates of air and liquid fuel are calculated for the chosen application and based on the type of fuel employed.
2. Dimensions of air ducts, capacities of air supply fans or compressors and sizes of slots or ports for air inlets into the combustion chamber are calculated.
3. Dimensions of fuel storage tanks, capacity of fuel pumps and sizes of hydraulic lines are estimated. A fuel injection method is also selected to ensure steady fuel feed rate.
4. Basic heat transfer calculations to estimate the heat loss and the thickness of the insulation required are carried out.
5. In wick burners, based on the rate of burning and the required flame dimensions, the sizes and number of the wicks required are calculated.
6. In pre-vaporizing burners, a fuel feed method and dimensions of the vaporizer tube to ensure sufficient heat transfer to steadily vaporize the fuel are selected.
7. In vaporizing burners, especially those involving sleeves, the annular gap between the sleeves, where the fuel is supplied, and the holes provided in the sleeves for air entry should be adequately designed based on the capacity of the burner.
8. In porous burners, the selection of the material and the porosity for the porous block is important. Depending upon the type of fuel used, the dimensions of the porous block are fixed.
9. In atomizing burners, depending upon the type of liquid fuel used, a suitable atomizer is chosen. Droplet sizes and size distributions are estimated for the chosen atomizer for the expected range of liquid supply rates. Based on the measured or calculated burning rate of an isolated mean sized droplet, the droplet life time is estimated. From the initial diameter and the initial relative velocity of the droplet, its penetration distance into the combustion chamber is estimated. Accurate estimation of the drag between the moving droplet and the ambient gas is required to calculate the penetration distance. The dimensions of the combustion chamber are fixed based on the spray angle and the maximum penetration distance, which a droplet can travel. Swirl-assisted air flow may be used to increase the residence time without increasing the length of combustion chambers, as otherwise the combustion chambers may have to be excessively long.
10. Peripherals to provide ignition source and to control the flame anchoring are suitably added.

In general, after carrying out the initial design calculations, a prototype burner is constructed and thorough testing is carried out to ensure the satisfactory performance of the proposed burner with the expected range of operating parameters. Numerical simulations may also be used to complement this exercise, in order to optimize several operating parameters, especially for atomizing burners.

5.5 Operation of Liquid Fuel Burners

A few important points to be considered during the operation of liquid fuel burners are summarized below.

Stable operation of wick burners depends on the type and purity of the fuel used, ventilation (movement of the surrounding atmospheric air) and the quality of the wicks used. Wick-based flames using kerosene type of fuels are luminous and are often accompanied by soot or smoke emissions. A notable amount of CO emission is also present when the aeration to these burners is insufficient. Use of volatile fuels such as alcohols and pine oil improves the combustion performance of wick burners.

Pre-vaporizing burners are sensitive to the mismatch in the rate of supply of liquid fuel and the rate at which the liquid is vaporized. Therefore, their operation range and turndown ratio are quite small. A controlled method of supplying the liquid fuel to the burner should be adopted to have steady burning features. Since atmospheric air entrains and mixes with the fuel vapor, stability and performance issues associated with an atmospheric entrained air gas burner are relevant here as well. In these burners, the gas phase emissions can be controlled by adjusting the air supply.

Vaporizing burners require the quality of the fuel to be high. Regular maintenance works such as cleaning the fuel line filters, air intake ports, side wall perforations and inner sleeve walls are required to ensure stable operation with low emissions. A steady supply of the required amount of air is essential in order to avoid soot emissions, production of CO and unburnt hydrocarbons which arise from the incomplete combustion of kerosene type fuels. Further, preheating the air used for combustion also helps in improving the performance. Burners having forced draft air supply are generally free from such emissions.

Performance of porous burners depends on the fuel type and its quality. An important factor in these burners is ensuring a steady fuel supply to the porous bed. Regular maintenance of porous bed to clean the carbon depositions within it is essential. There is a possibility of nitric oxide formation in these burners, if the temperature of the porous bed increases above a certain value. Cooling the porous bed, adding diluents and recirculation of small amounts of product gases reduce the possibility of nitric oxide emissions.

Performance of atomizing burners depends upon the selection of a suitable atomizer for a given application. In general, preheating the liquid fuel results in a better spray quality. In these burners, the design of the combustion chamber is an important factor in improving the combustion efficiency.

Pressure atomizers require scheduled cleaning of the orifices and fuel lines. Liquid fuel is fed at the designated supply pressure throughout the operation and any perturbations in this will result in a bad quality spray and incomplete burning of the liquid fuel.

Rotary burners need finer adjustments in order to ensure a steady operation. These burners are oriented vertically or horizontally, based on the requirement. Heavier liquid fuels are burnt in horizontal rotary burners. The rotating surface and the air nozzles have to be cleaned almost on a daily basis to remove any carbon deposits in them, so as to improve the performance and emission characteristics. Failure to do so will invariably result in smoke emissions and perhaps even a fire.

High pressure air-assist burners require a proper supply of primary and secondary air to ensure complete combustion of the spray droplets. The atomization air, which is a part of the primary air, is used to control the spray quality, and the secondary air is used to control the combustion efficiency and the emission levels. Multi-zone air supply is also provided where required. In low pressure air-blast burners, primary air flow passages should be designed and fabricated with ultimate care, since the primary air sent through the atomizer controls the spray quality and also enhances the mixing of fuel vapor and air. Secondary air supply ensures complete combustion.

Oil burners are used quite extensively in domestic and industrial applications. The main challenge in these burners is in using different types of fuels as well as blends of different fuels without any adverse effect on the performance and emission characteristics. Hybrid burners such as low pressure air-assist burners, which operate using gaseous and liquid fuels, either individually or in combination, are used in several industrial applications. Burners with high turndown ratio are also necessary for several applications.

Review Questions

1. List the types of liquid fuel burners used in domestic and industrial applications.
2. Briefly discuss the mechanism of operation of a wick stove.
3. How the performance of a pre-vaporizing burner is controlled?
4. What are the advantages and disadvantages of a vaporizing burner?
5. Define Weber number and Reynolds number.
6. Explain with clear sketches the regimes of atomization and the range of Weber numbers associated with those regimes.
7. List the different types of atomizers used.
8. Discuss the differences between duplex and dual orifice atomizers.
9. Explain the mechanism of atomization in a rotary atomizer with flat rotating surface.
10. List the differences between internal mixing and external mixing twin fluid atomizers.
11. What is Sauter mean diameter?
12. List the important processes involved in spray combustion.

Exercise Problems

1. In an atmospheric pressure pre-vaporizer, a horizontal porous wick, saturated with methanol, is kept in flush with the bottom wall of a horizontal wind tunnel. Air flows at a given velocity over the surface of the wick at a temperature of

320 K. As a result, methanol vaporizes and mixes with air and the reactant mixture flows out of the wind tunnel into the combustion chamber. The cross section of the wind tunnel is 0.1 m × 0.1 m. Surface area of the porous plate is 0.05 m². At an air velocity of 0.5 m/s, the evaporation rate per unit area of the porous plate is 0.017 kg/m² s and when the air velocity is increased to 2.5 m/s, the evaporation rate increases to 0.027 kg/m² s. Determine the equivalent ratio of the reactant mixture at these air velocities.

2. An atmospheric pre-vaporizer burner has a power rating of 5 kW. It operates with kerosene (boiling point 225 °C, latent heat of vaporization 250 kJ/kg and specific heat 2 kJ/kg-K), which is fed at a temperature of 25 °C. Its calorific value is 43 MJ/kg and the average heat flux from the flame to the tube carrying the liquid fuel is 60 kW/m². If the tube is made of copper and has a diameter of 10 mm, determine the length of the tube to pre-vaporize kerosene.

3. Density of canola oil varies with temperature (in °C) as $926.7-0.606\,T$ kg/m³, its viscosity varies as $8.3\,T^{-1.52}$ Pa-s and its surface tension varies as $0.0326\,T^{-0.05}$ N/m. The temperature range valid for these correlations is 25–200 °C. If the oil is heated to a given temperature and injected using an atomizer with orifice diameter of 0.5 mm at 2 bar pressure, determine the regime of atomization for oil temperatures of (a) 50 °C, (b) 100 °C and (c) 200 °C.

4. A liquid fuel droplet of 2 mm diameter is moving in quiescent air. Ambient temperature is 300 K and 1 atm pressure is maintained. Calculate the minimum Weber number at which droplet breaks.

5. Consider a n-heptane droplet with an initial diameter of 1 mm injected into a furnace having stagnant nitrogen at 500 K and 1 atm. The initial droplet velocity is 10 m/s. The drag force experienced by the droplet, $F = \rho\,C_D\,(V^2/2)\,(\pi d^2/4)$. Here, ρ is gas density, d is droplet diameter, C_D is drag coefficient and V is droplet velocity. This drag force varies with time as the droplet decelerates, $F = m_d(dV/dt)$, where m_d is the mass of the droplet. Drag coefficient, C_D, is determined by a correlation given as $C_D = 24/Re_d$. Here, Re_d is the Reynolds number calculated using droplet diameter, its velocity and kinematic viscosity of the mixture. The evaporation rate of the droplet in a convective environment can be determined using, $\dot{m} = 2\pi\,(\lambda/c_p)\,(d/2)\,Nu \times \ln(1 + B_T)$. Here, Nu is Nusselt number calculated as, $Nu = 2 + 0.6Re_d^{0.5}Pr^{1/3}$, where Pr is Prandtl number. Determine the variation of d and V with time and the distance the droplet travels when its diameter reached 1% of the initial value. Values of specific heat at constant pressure and thermal conductivity may be taken as 4390 J/kg–K and 0.111 W/m–K, respectively. Other properties may be taken from NIST database.

Chapter 6
Solid Fuel Systems

Solid fuels were introduced in Chap. 1. In this chapter, aspects relating to the combustion of solid fuels are discussed. Commonly used solid fuels such as coal and biomass are heterogeneous materials. They contain trapped *moisture*, *volatile* matter, *fixed carbon* and *ash*, which are determined by proximate analysis. Coal is perhaps the most widely used fuel around the world for the generation of electricity. It is also abundantly available in many countries around the globe. However, it is quite difficult fuel to use in terms of handling, performance, emissions and ash removal. The type of burner employed to burn coal depends upon its composition and physical properties. Therefore, a burner designed to burn a particular type of coal, in general will not perform as well for other types of coals. These points are true for different types of biomass fuels as well.

6.1 Combustion of Solid Fuels [2, 20]

Solid fuels need a preparation process before they are burnt in a combustion chamber. In the preparation process, solid fuel blocks are first crushed into particles. Further, they are segregated, washed and palletized, depending upon their source and nature. For example, solid fuels with high ash content are washed in order to reduce their ash percentage. Biomass fuels such as saw dust, rice husk and wheat husk are powdered and palletized into small chunks of different shapes. Based on the rate of firing, the solid fuel particles are oxidized either by packing them on a grate which may be stationary or moving, or by suspending them in air. To burn solid fuels on grates, particles with sizes greater than 6 mm are fed into the combustion chamber either at the top or the bottom of the fuel bed. They burn forming different reactive layers within the bed. On the other hand, in suspension firing, particles having sizes less than 6 mm are fed into the combustion chamber using different techniques such as screw-conveying and air-assisted feeding. The particles are suspended in the air and

© The Author(s), under exclusive license to Springer Nature Switzerland AG 2022
V. Raghavan, *Combustion Technology*,
https://doi.org/10.1007/978-3-030-74621-6_6

mix with air, depending on their size distribution and air velocities. The combustion of coal and other solid fuels is carried out in four main stages:

(1)   *drying* of moisture and *pyrolysis* or release of the volatile matter (*devolatilization*),

(2)   *ignition* and *heterogeneous burning* of the fixed carbon resulting in the formation of carbon-monoxide and carbon dioxide,

(3)   completion of *homogeneous* (gas phase) reactions, which includes the oxidation of the volatiles and other fuel gases, formed during pyrolysis,

(4)   satisfactory *disposal of the ash* with negligible carbon content in it.

These four stages are commonly seen in all the solid fuel burners. However, the sequence of these stages and the rate at which the fuel enters and leaves each of these stages vary with different burners. The stages may even overlap in some burners.

Drying and pyrolysis are thermal decomposition processes. Heating of the fuel particles is accomplished by convection in suspension firing and mostly by radiation in grate firing. When the particles are heated to more than 100 °C, moisture is released. Devolatilization of volatile matters by chemical cracking of the compounds in the solid fuel starts around 300 °C. This is followed by the formation of tar, which is a liquid at normal temperatures, and gaseous products. Tars are organic compounds having a structure same as their base fuel. They evaporate between 500 and 600 °C and condense at lower temperatures. In general, the particle shape remains unchanged up to 400 °C, and it begins to soften as the temperature increases above this value. Semi-coke is formed as the volatiles are released. Semi-coke is an amorphous solid with distinct pores and cavities and an enlarged surface. Further devolatilization occurs and only fixed carbon with ash, called coke, remains in the particles. The percentage, composition of volatile compounds and their release rate depend on the type of solid fuel, heating rate in the furnace, particle size and the temperature inside the furnace. Ignition of coke particle occurs when its temperature is greater than 800 °C. In several circumstances, volatiles mixed with air are ignited and this heats up the coke particle. Combustion of coke (carbon) occurs in a heterogeneous mode as discussed in Chap. 3. Volatiles and other fuel gases such as CO burn in a homogeneous mode. The burn-out of coke is the rate-limiting step in coal combustion. Sufficient residence time must be provided for the solid fuel particle inside the furnace to accomplish complete burn-out. The burner construction should provide an easy access to remove the ash that is left behind. Ideally, the combustion process should produce ash with negligible carbon content in it.

6.1.1   Grate Burners, Fluidized Beds and Pulverized Burners [2, 16, 20]

In this section, details of different types of solid fuel burners are presented. Depending upon the power requirement, the rate of firing of a solid fuel, having a given heating value, is determined and a proper burner is selected. The heating value of a solid

Table 6.1 Typical features of different solid fuel burners

Burner	Heat release rate (MW/m^2)	Particle size (mm)	Height of the fuel firing zone (m)	Temperature range (°C)
Grate burners	0.25–1.5	50–6	0.2–0.3	1200–1500
Bubbling fluidized bed burners	0.5–1.5	<6	1–2	900–1100
Circulating fluidized bed burners	3–5	<6	10–30	850–1000
Pulverized burners	4–6	<0.1	25–50	1600–1800

fuel is obtained from its ultimate analysis, where the fractions of elements such as C, H, S, N and O are obtained on a mass basis. For example, Dulong's formula for evaluating the heating value (*HV*) in MJ/kg is,

$$HV = 33.82\,C + 141.79(H - O/8) + 9.42\,S$$

Every burner is designed to provide the necessary residence time for completely burning the solid fuel particles. The burner dimensions, the size range of the solid fuel particles and the oxidizer flow rate are fixed accordingly. Size of the solid particles, which usually are not spherical, is generally determined by allowing the particles to pass through several *sieves* of different hole sizes. The sieves are vibrated by a sieve-shaker. The particles are classified based on their size into four groups A, B, C and D by Geldart. Group C particles have the smallest size (<20 μm), followed by group A (20–90 μm), group B (90–600 μm) and group D (>600 μm). In order to ensure complete combustion, excess air is almost always supplied. Table 6.1 presents the typical features of commonly used solid fuel burners.

Grate burners are the simplest in terms of construction and operation. These have an oxidizer supply system, which is primarily a distributer plate or a grate having several holes through the bottom of which the oxidizer is supplied. In general, the solid fuel particles are fed on to the top of the grate. The fuel feeding is accomplished by different methods. In a few cases, the solid fuels are also fed from the bottom of the grate. Based on the type of the solid fuel used, several configurations of grate burners are available. From Table 6.1, it is clear that the height of the fuel firing zone is the least for the grate burners, which handle solid fuel particles of larger sizes. However, the bed area (diameter of the burner) is relatively large for a given power output, since the flux of the heat release rate is quite small. Therefore, grate burners are generally limited to 200 MW of thermal power output. Although it is possible in principle to increase the power rating, the installation costs become quite high. The advantages of grate burners are their simple construction, easy operation and minimum crushing charges, as larger particles are used. The operating temperatures are in the range of 1200–1500 °C. At these temperatures, the ash is quite likely to melt and the furnace must be designed keeping this in mind and provide for the removal of molten ash. Similarly, particles of size less than 6 mm cannot be efficiently burnt in these burners.

Fluidized bed burners are classified as bubbling and circulating fluidized bed burners. In these burners, the oxidizer is fed through the bottom of a chamber where combustion occurs. The area of cross section of the chamber (usually called a riser column) and its height are decided based on the type of solid particles and their firing rate. Solid fuels are fed from different locations on to a distributer plate (akin to a grate) through which oxidizer is supplied. The velocity of the oxidizer is high enough to suspend particles smaller than 6 mm. In bubbling fluidized bed burners, as the oxidizer flows through the initially packed bed of particles, the bed expands and particles transition to a suspended state due to the drag force exerted by the gas. The oxidizer bubbles through the particles and also carries several particles along with it. The particles are carried to a certain height and eventually fall down, when the drag force exerted by the gas is no longer sufficient enough to carry them. This bubbling action continues and provides an adequate residence time for the particles to burn completely. Only very tiny particles, predominantly ash, are carried away along with the flue gases.

On the other hand, the circulating fluidized bed has a much higher oxidizer (and flue gas) velocity. The particles are carried away by the oxidizer and reach the top of the riser column. The particles carried by the gas mixture pass through a cyclone, where most of the particles are separated from the gases due to centrifugal action. The separated particles are fed back to the burner, and this process continues for a few cycles ensuring sufficient residence time for complete combustion. Only very small particles, which are usually ash particles, leave through the cyclone along with flue gases. Bubbling fluidized beds have almost the same bed area as that of grate burners for a given power rating. However, bubbling beds are taller than grate burners and the combustion efficiency is better in fluidized beds as the smaller particles are convectively burnt in them. The operating temperature is also lower for a bubbling bed than that of a grate burner. Therefore, usually the ash does not melt and the solid ash particles are easily removed. Circulating fluidized beds have a smaller bed area than bubbling beds for the same power rating to facilitate higher oxidizer velocity. However, the combustion zone in circulating beds is taller than that of a bubbling bed. They have additional components such as cyclone (to separate the particles) and loop seal (to feed the particles into the bed where the oxidizer is supplied). Their operating temperature is slightly less than that of a bubbling bed. Circulating fluidized beds are used in installations delivering up to 800 MW (thermal).

Pulverized burners handle micron-sized particles. In the vertical configuration, they are taller than the circulating fluidized beds. Due to the higher firing rates involved in these burners, the gas velocities, heat release rate per unit area and operating temperature are higher when compared to other burners. They are employed in larger installations which deliver up to 2500 MW (thermal). Several configurations of pulverized burners are available for firing the solid fuel particles. Some important features of the burners discussed above are presented in the subsequent sections.

Table 6.2 Advantages and disadvantages of grate, fluidized bed and pulverized burners

Grate burners	Fluidized bed burners	Pulverized burners
Advantages: 1. Economical fuel preparation 2. Easier design and construction 3. Easier and simpler operation 4. Less auxiliary power requirement 5. Low nitric oxides emission 6. Easier removal of sulfur by adding limestone *Disadvantages*: 1. Ash trapped unburnt carbon up to 2–4% 2. High flue gas temperature 3. Not suitable for particles smaller than 6 mm	*Advantages*: 1. Economical fuel preparation compared to pulverized burners 2. Better combustion efficiency and power control 3. Easy flue gas cleanup *Disadvantages*: 1. Higher limestone demand when excess air is used (Bubbling beds require lower limestone than circulating fluidized beds) 2. Ash not usable as it is and requires further processing 3. Circulating fluidized beds suffer from erosion problems	*Advantages*: 1. Large capacity installations possible 2. High power density (the plant volume is much smaller for a given power rating) 3. Good combustion efficiency 4. Usable ash *Disadvantages*: 1. Very high fuel preparation cost involved 2. Costly flue gas cleaning process

Table 6.2 presents a comparison of advantages and disadvantages of grate, fluidized bed and pulverized burners. In general, grate burners are ideal for small-scale installations, where the cost of installation and operation can be quite low. When the ash content in the coal increases, ash contained unburnt carbon level may increase even up to 4%. The ash is generally used in several other applications including fertilizer and cement industries.

Fluidized bed burners have high combustion efficiencies and offer robust control of operation. But the installation and operation costs are higher than grate burners. Bubbling bed burners are better than circulating fluidized bed burners in terms of erosion control in inline heat exchangers. Pulverized burners are suitable for large installations, and good combustion occurs in these burners. However, the fuel preparation cost is usually quite high.

6.1.2 Design, Performance and Emission Characteristics

Grate Burners

The simplest type of grate burner is the overfed fixed bed type. This is generally used for burning coking fuels, where the fixed carbon content is high and the coal particle transforms to a porous coke particle when heated to more than 1000 °C in an inert environment. Examples of coking fuels are anthracite and low volatile bituminous coals. Here, the raw fuel is supplied to the fuel bed from the top, where the flue gases leave. The fuel is fed in small quantities at frequent intervals and it goes

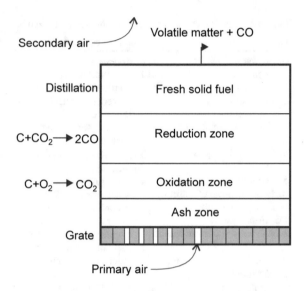

Fig. 6.1 Schematic of zones of reaction involved in overfed fixed bed burner

through various phases as schematically shown in Fig. 6.1. As the fresh fuel particles are fed from the top, they are heated up by the intense radiative heat from below. As the particles are heated up, the trapped moisture evaporates and eventually the volatiles are also released. The regime where these happen is generally termed as the distillation zone. The volatiles are burnt completely using the secondary air supply. Only coke (fixed carbon and ash) remains to form the subsequent lower zones in the bed. The layer next to the distillation layer is called the reduction zone, where the surface reaction between the hot carbon particles and carbon dioxide takes place.

This heterogeneous reduction reaction, which is an endothermic reaction, may be written as,

$$C(s) + CO_2 \rightarrow 2CO$$

The CO formed in the reduction zone is oxidized by the secondary air stream supplied at the top of the bed. Carbon dioxide is formed in the layer below the reduction zone as a result of the oxidation of the hot coke particles with the air that enters from the bottom of the fuel bed through a distributer plate, as shown in Fig. 6.1. The heterogeneous carbon oxidation reaction, which is exothermic in nature, may be expressed as,

$$C(s) + O_2 \rightarrow CO_2$$

The entire carbon in the coke is burnt and the ash settles as the bottom-most layer. The ash is frequently removed in molten form or in solid form. The porous layer of ash over the grate and below the oxidation zone helps to distribute the flow of primary air into the combustion zone. The thickness of the ash layer is controlled by a planned removal of the ash from the bed. The thickness of the each zone depends on the type of solid fuel and average size of the particles used. The combustion gases, consisting of carbon monoxide and carbon dioxide, and nitrogen, pass through the distillation zone and mix with the volatile matter and moisture. The unburnt fuel gases in this mixture are burnt completely using the secondary air, which comes in as a turbulent stream usually. If the secondary air does not mix with the gas mixture containing volatiles, then the tar released from the volatiles may lead to soot formation and release of blackish smoke. Thus, the supply of secondary air is important to avoid pollutant formation. Figure 6.2 illustrates the typical variation in gas composition in the ash, oxidation and reduction zones, as a function of the height above the grate in an overfed fuel bed. The percentage of the gases may vary depending on the fuel type.

In these burners, the open spaces or voids in the coal bed facilitates the escape of the oxygen in the primary air through the fuel bed. Therefore, the fuel has to be fed carefully to avoid such leaks. The burning rate in the coke bed is limited predominantly by the rate with which oxygen is transported to the surface of the hot carbon. In other words, in these burners, the burning rates are proportional to the flow rate of the primary air, which is a function of the pressure difference across the fuel bed. The primary air flow rate is controlled by raising the pressure beneath the fuel bed. Depending upon the type of the solid fuel used, the thickness of the oxidation layer may be estimated. The fuel is fed at a rate such that this thickness is maintained under steady operation.

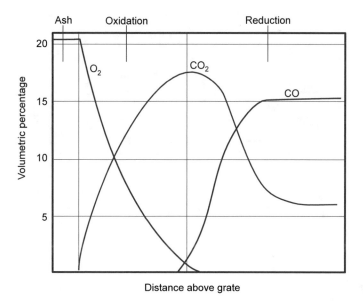

Fig. 6.2 Typical distribution of gases inside the overfed grate burner

For solid fuels having little volatile matter such as anthracite coal, the fuel particles are fed evenly over the thin incandescent coke surface, so that the particles are ignited by the intense heat of the bed. On the other hand, solid fuels with a large amount of volatile matter, such as high-volatile bituminous coal and biomass, require a different method of firing. Such fuel particles are fed in such a manner that the raw fuel is intermingled between the incandescent coke particles. As a result, the volatile matter coming out of the particles, upon mixing with the air, is ignited by the intense heat of the hot coke.

There are variety of grate burners where the grate is mechanically moved so as to physically transport the raw fuel from one side of the burner to the other. Such burners are also called *stokers*. Coal is fed from a hopper onto a grate that moves through the combustion chamber as shown in Fig. 6.3. This type of burner is also an overfed burner. The moving grate is flexible flat surface, composed generally of a number of cast-iron links. The grate is turned over by sprockets at each end like that in a conveyor belt. The interspacing between the links forms the passage for air supply from the bottom. This interlink distance can be adjusted to support particles with a certain size range. Stokers of this type can be built in sizes up to around 8 m wide by 12 m long. In general, firing rates of around 70–150 kg of coal per square meter of grate per hour are possible in these burners. The moving grate is driven usually by a motor or an engine, through a gearbox to vary the speed of the movement of the grate. Air for combustion enters from a *plenum chamber* or *wind box* beneath the grate. In general, grates having forced air supply are designed to

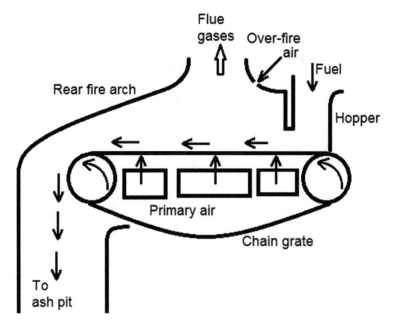

Fig. 6.3 Schematic of a moving grate burner

have 5–10% of the total surface area as holes to allow adequate air flow, however, without allowing fine fuel particles to escape. For natural draft, nearly 20% of the grate area may be required for air flow. The air space beneath the grate (wind box) is divided into a number of compartments to have variable control on the air fed into the combustion chamber at different positions on the grate. The grates also have interlocking side plate to prevent air loss around the sides. The passage to the ash pit is also sealed properly to minimize the loss of air into the ash chamber. Grate stokers are well adapted for burning anthracite, semi-anthracite, non-caking bituminous (non-caking fuels leave powdery residue after volatiles are released), sub-bituminous and lignite types of coals. The fuel particles go through several phases as they travel on the grate. First, moisture and volatiles are released and coke particles are formed. Next, the fixed carbon in the coke undergoes complete combustion in the presence of the primary air that is supplied. This leaves behind ash particle on the grate, that needs to be disposed of. Caking coals, such as caking bituminous, cannot be burnt on these burners. Caking coals leave chunks of porous solid residue when moisture and volatiles are released. Such lumps of coke particles on the grate obstruct the flow of air resulting in incomplete combustion. The size of the particles should be well-controlled in these burners and fine particles should be removed. In a few cases, when bituminous coal is used, water is sprayed on the particles in order to increase the moisture content in the coal to 12–14%. This helps in regulating the burning rate and brings down the carbon losses.

Ignition of the fresh fuel is accomplished by a combination of three factors: (1) radiation from an hot arch-wall made of fire brick; (2) hot gases flowing over the surface of the raw coal; and (3) fine incandescent coke particles, which bubble due to the air flow, settling on the raw fuel surface. The coke particles receive intense heat from the hot rear-arch wall, as shown in Fig. 6.3. This is a major factor in promoting instantaneous ignition. This type of arch wall is constructed in the front portion of the grate or on both sides as well. Front arch is best suited for high volatile bituminous coals as the volatiles released mix with the air and the reactant mixture is ignited by the radiation from the hot wall. The arch shape also helps in controlling the flow of gases toward the unburnt coal surface. For example, the rear-arch creates a high gas velocity zone near the exit and this helps in transporting the fine incandescent particles back to the zone where the raw fuel is fed. Near the entry to the furnace, the gases contain primarily the volatiles released from the fuel particles. Near the exit, the fuel bed is thin, and the gases above consist mostly the excess air. Turbulent mixing of these gases is necessary to ensure complete combustion and this may be partially accomplished with a good arch design. Steam and air jets are injected at some height in the furnace arch to achieve smoke-free, complete and efficient combustion. The ash particles, which settle toward the bottom of the bed, protect the grate surface from excessive heat. Since the grate is not agitated by any mechanical means, this leads to clinker formation and operational problems.

Inclined grate stoker burners are used to burn a variety of solid fuels. A schematic view of such a burner is shown in Fig. 6.4. This is also an overfed stoker. Here, the fresh fuel is pushed on to a horizontal grate platform, where it ignites due to the heat of the fuel burning below in the inclined grate. When the next charge of raw fuel

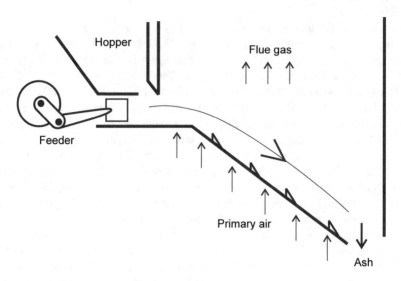

Fig. 6.4 Schematic of an inclined stoker grate burner

is pushed into the furnace, the ignited particles are pushed on to the inclined grate. The particles gradually descend along the inclined grate to the bottom. During the travel down the grate, the lumps of the fuel particles are broken up thus ensuring that complete combustion occurs. At the bottom, the ash particles are dumped into the ash pit. Protrusions, called lugs, are installed at several locations in the inclined grate. These help in holding the fuel on the grate and prevent the burning particles from falling into the refuse pit before all the fuel particles are completely burnt. The inclined grate is usually limited in size to small units. Refuse fuels such as wood bark, lignite and wood pulp are commonly burnt in this burner with rather high excess air in small installations, as there are chances of unburnt fuel being dumped into the ash pit.

Another variant of the overfed stoker burner is the spreader stoker burner. In this the fuel is fed in such a way that most of the fuel particles are oxidized while they travel in a suspended mode in the hot furnace and the remaining fuel particles are burnt after they reach a stationary fuel bed. A schematic diagram of such a burner is shown in Fig. 6.5. The spreader stoker burner consists of a fuel hopper, feeder, distributor and horizontal stationary grate. Solid fuel from the hopper is metered by the feeder, which is a small conveyor belt that carries the fuel particles. This feeds the fuel to the distributor at a desired steady rate. A distributor, in its simplest form, consists of a paddle wheel with blades. The blades are angled such that they throw the particles in different trajectories across the stationary grate. Larger particles are thrown a longer distance, while the finer particles are dropped at the front of the grate itself. If uniform sized particles are used, then they tend to collect around one location in the burner. Large particles fail to burn completely and contribute to high carbon loss in the ash pit. On the other hand, an excessive number of fine

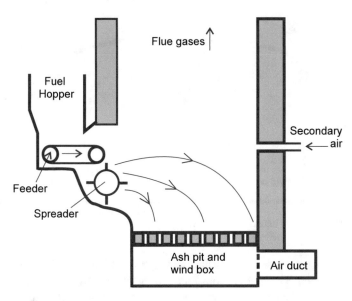

Fig. 6.5 Schematic representation of a spreader stoker burner

particles also contributes to carbon loss because they are carried away by the gases up the stack. Therefore, a proper particle size distribution is required in these burners. In some installations, pneumatic feeding nozzles are used to feed the fuel particles. Ignition of the particles occurs almost instantaneously from the intense heat radiation of the hot fuel bed and also due to the high temperature combustion gases. Volatiles are released, ignited and burnt simultaneously. The fixed carbon is also ignited and partially burnt before the hot particles settle down on the fuel bed.

The grates are made in several designs and may be stationary or moving. The space below the grate is used as the ash pit. The coke zone on the fuel bed is very thin and after the carbon is burnt completely, the ash layer may build up to a depth of around 100 mm, before it is dumped. The ash layer prevents the grate from overheating. The air spaces in the grate occupy only about 5% of the total grate area to minimize the carbon losses. This also facilitates the creation of air jets through several holes, which may be distributed almost uniformly over the fuel bed. Secondary air is added to the furnace through openings above and below the spreader. In several cases, better efficiency is obtained by forcing most of the air through the fuel bed. Steam may also be supplied. The capacity of this burner may be increased by adding more feeders in parallel and making the grate and furnace wider. The changes in the feeding rates are handled by operating only the required number of feeders. Any grade of coal starting from semi-anthracite to lignite may be burnt in these burners.

A sketch of an underfed fuel bed is pictorially shown in Fig. 6.6. Fresh fuel particles are fed into the bed from below using a power ram or a screw conveyor. The raw coal below the combustion zone is heated by the intense heat from the burning fuel bed. The volatiles are released as the coal is forced into the active burning zone,

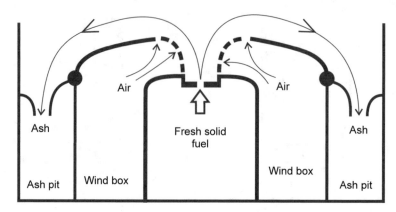

Fig. 6.6 Schematic of an underfeed grate burner

and they are heated to a sufficiently high temperature to cause ignition, which occurs just above the bed. The grate is constructed to be sloped toward the sides, and the ash formed above the bed of coke is moved toward the edges and removed.

Air enters the fuel bed through openings in the cast-iron supports of the fuel bed. Most of the volatiles mix with the primary air and burn as the mixture flows through the coke bed. The lower part of the coke bed serves as the oxidation zone. A reduction zone, in which the carbon reacts with carbon dioxide, is present above this layer just as in the case of overfed burners. Secondary air is supplied above the bed, and this aids in the completion of the burning of the combustible gases rising above the fuel bed. High turbulent flow of secondary air is not required in these burners.

Fluidized bed burners

In grate burners, solid fuel particles with sizes larger than 6 mm are fed on to a distributor plate to form a *packed bed*. The oxidizer supplied beneath the distributor plate flows through the *pores* or *gaps* or *voids* between the solid particles. For a given rate of solid fuel supply, the rate of oxidizer supply is usually fixed considering the excess air requirement. In grate burners, this oxidizer flow is not capable of fluidizing the large particles present over the bed. The packed bed of particles causes a certain pressure drop depending upon the oxidizer flow rate, and the supply pressure of the oxidizer is adjusted to ensure the required flow of oxidizer into the furnace. The pressure drop increases as the oxidizer flow rate is increased. On the other hand, when fine-enough solid particles encounter a steadily flowing oxidizer stream having sufficient momentum, the initially packed bed of particles expands and eventually transforms into a *suspended* state. The weight of the solid particles is overcome by the drag force exerted by the oxidizer flow. Depending upon the fineness of the particles and the momentum of the gas stream, the particle bed expands up to a certain height. The particles are not carried by the gas beyond this height. The gas–solid particle mixture in the expanded bed exhibits a *liquid-like* behavior. Characteristics such as static pressure varying as a function of the height of the bed, bed surface maintained horizontally, lighter particles floating over the bed and denser particles sinking into the bed are observed. Based on the volumetric flow rate of the oxidizer and the cross-sectional area of the bed, an average velocity, termed as *superficial* velocity, is calculated. When the superficial velocity is just enough to expand a packed bed, it is termed as the *minimum fluidization velocity*. When the superficial velocity is increased beyond the minimum fluidization velocity, the pressure drop across the bed remains almost a constant. Based on the particle size distribution, the fluidized region acts as a *well-stirred* column, having almost uniform properties throughout the column. As the gas velocity is increased, the expanded height of the particles increases. *Bubbles* are formed from the gas inlets and they move upward carrying some amount of solid particles in their wake region. The solid–gas mixture outside the bubbles is called an *emulsion*. The bubbles reach the top of the expanded bed height and burst, throwing the particles upward, most of which eventually fall down. This regime is called the *bubbling fluidization* regime. When the oxidizer flow rate is further increased, the superficial velocity increases and more number of bubbles are observed within the bed. At an even higher superficial velocity, the bubbles are observed to rapidly break and/or coalesce with other bubbles. This vigorous bubble dynamics occurs in the *turbulent fluidization* regime. For burners operating in this regime, in general, a cyclone separator is installed to capture the particles escaping along with the gases. *Terminal* velocity is the equilibrium velocity attained by a particle, freely falling under normal gravity in the atmosphere. When the superficial velocity is increased beyond the terminal velocity, most of the solid particles are carried away by the gas flow. The particles are captured using a cyclone and fed back into the bed using a loop seal. This is termed as *circulating fluidization* regime. While a cyclone separates the particles using the centrifugal force, a loop-seal is used to locally increase the pressure of the gas–solid mixture so that it can be fed back in to the dense bed. The bubbling and circulating fluidized bed burners are illustrated in Fig. 6.7.

The height of the combustion chamber in fluidized bed burners is an important parameter, especially for the bubbling fluidized bed burners, since they have free space above the expanded bed height, called the *freeboard*. Bubbles break on the surface of the bed and particles are thrown into or entrained by the gases flowing upward into the freeboard. They travel to some height and fall back when the drag force exerted by the gas flow less than the weight of the particles. The freeboard extends up to a certain height called transport disengagement height (TDH). Only very tiny particles are carried beyond this point. The height of the combustion chamber is more than the TDH. The hydrodynamics of bubbling fluidized bed is schematically shown in Fig. 6.8. To control the size and growth of bubbles, fluidized beds are also operated at pressures higher than atmospheric pressure.

The temperature field in a fluidized bed is quite uniform. This is due to the presence of well-distributed inert particles of ash and bed material, sand. A good amount of heat transfer occurs between the solid fuel and the inert particles, as well as between the gas and the particles. As mentioned earlier, the entire bed behaves as a well-stirred reactor consisting of gases, solid fuel and inert particles. As a result, the operating temperature is usually in the range of 800–950 °C. Due to the convective heat transfer

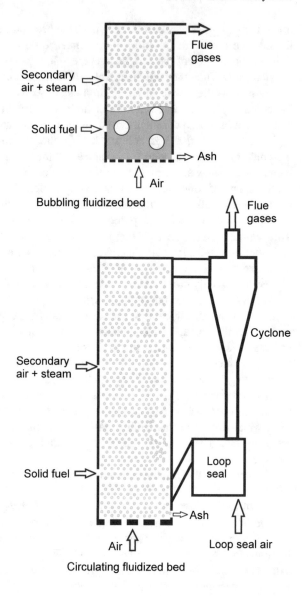

Fig. 6.7 Schematic representation of bubbling and circulating fluidized bed burners

in the fluidized bed, the temperature of the raw solid fuel particle increases at a rapid rate. Moisture is released when the particle temperature is just more than 100 °C. Volatiles in the fuel particles are released when the temperature reaches around 400–450 °C. The release of moisture as well as volatiles may cause fragmentation of the particle, if the pores in the solid particles are not large enough for the vapor and gases to escape. Due to continuous heating, the gas inside the solid particle may be pressurized causing the particle to break. When fragmentation occurs, the fluidization

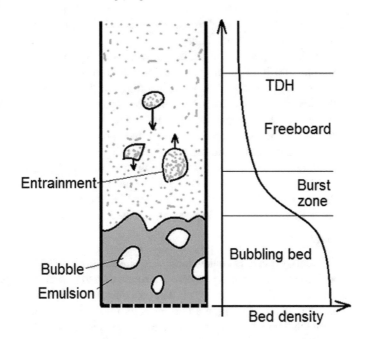

Fig. 6.8 Pictorial representation of hydrodynamics of a bubbling fluidized bed

conditions, bubble dynamics, heat and mass transfer characteristics may change in the bed.

In general, under most operating conditions, the volatiles burn in the freeboard. Therefore, it should be ensured that enough oxygen reaches the freeboard. Simultaneously, ignition and combustion of char occur in the bed. The internal structure of char is significantly different from the raw fuel. Combustion of the char particle takes place at its surface and in its pores. The transport of oxygen toward the char surface and its penetration inside the char depend upon the mass transfer characteristics between the bubble and the emulsion. While the carbon in the char burns, if the ash separates out from the char particle, then the size of the char particle decreases almost steadily as it burns. On the other hand, if the ash does not separate while the carbon burns, a porous ash particle of almost the same size as the char particles remains. In the bubbling or in the burst zones, depending upon the solid fuel, attrition of the particles takes place as a result of collisions between the inert and the char particles. The resultant small char particles may not burn completely and may flow out of the combustion chamber along with the gases.

The rate of release of volatiles and its burning affect the overall performance of the combustor. If large amounts of volatile gases are released in the dense bed, the mixing of the gases with oxygen and its subsequent burning should be ensured. Despite the supply of secondary air in the freeboard and its intense mixing with volatiles, some unburnt fuel may leave the combustion chamber if sufficient reaction time is not available within the freeboard. Therefore, in several installations the diameter of the freeboard is larger than the bubbling zone.

In fluidized beds, it is usual to operate the burners at an average temperature in the range of 800–900 °C. Recalling the discussions from Chap. 3 regarding the combustion of a carbon particle, at these temperatures, it is reasonable enough to assume that the char burning occurs in the diffusion regime. However, this further depends upon the particle size, porosity and the reactivity of the solid fuel. Off-design operation of the bed, in general, results in incomplete combustion owing to reduced availability of oxygen. Fluidized beds are sensitive to operating parameters such as temperature and pressure. Proper and vigorous fluidization, uniform fuel distribution, good mixing of volatiles in the bed and intensive interaction of bubbles with the emulsion phase are achieved by employing higher fluidization velocity, by producing smaller bubbles and by using proper particle sizes and distribution.

Pulverized burners

Burning of solid fuels, especially coal, in large-scale installations is carried out using pulverized fuel burners. The coal is powdered to micron-sized particles in a pulverizing mill and mixed with heated primary air. The coal dust and air mixture is supplied into the combustion chamber through a port, directly into a live flame. Due to the intense heat from the flame, the fine coal particles are ignited. Almost instantaneous release of moisture and volatiles take place and volatiles and char start to burn. In order to ensure that the ignition, pyrolysis and combustion are rapid, particles should be in the size range of 50–75 μm. In general, pulverization is accomplished in a ball-mill or an impact-mill. A ball-mill consists of a horizontal cylinder, the inside of which is partially filled with steel balls having 25–50 mm diameter. This cylinder rotates on a roller support at a given speed. Fresh coal and air are supplied through one end of the cylinder. As the cylinder rotates, the balls move randomly crushing the internally mixed coal particles. Heated air flowing through the cylinder entrains the fine coal particles and carries them away from the mill. The primary air with coal fines passes through certain sieves, where the oversized particles are filtered out. Only the primary air and fine coal particles in the size range of 50–75 μm are admitted into the combustion chamber. The filtered particles are again fed into the mill for further crushing.

A schematic representation of a horizontal pulverized burner is shown in Fig. 6.9. Primary air and pulverized coal are delivered to the burner through a pipe and are injected from a central nozzle. Small directing vanes in the nozzle impart a circumferential flow component to the primary air–coal dust mixture leaving the burner. This ensures a uniform mixture of air and coal dust at the burner exit. Secondary air from the air chamber is allowed to flow through an annular space surrounding the central nozzle. The secondary air passage is also installed with vanes in order to impart a turbulent swirl component to the secondary air flow. Turbulence level, flame shape and its extents in the combustor can be adjusted by controlling the secondary air flow rate and the angle of guide vanes. Ignition of pulverized coal dust–air mixture is accomplished by a fuel gas or oil torch, located coaxially within the central nozzle.

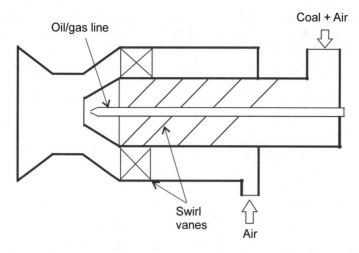

Fig. 6.9 Schematic of a horizontal pulverized burner

In this manner, it is possible to operate this burner with coal, fuel oil or gas and combination of these fuels.

In practice, it has been observed that the quickest ignition is obtained with a rich mixture of coal dust and air. The rate of ignition increases as the percentage of volatiles in the coal increases and the ash content decreases. Further, the high-volatile content fuels are observed to be inflammable over a wider range of air–fuel ratios when compared to the high ranked coals with high carbon content. The following trends have been seen in pulverized coal combustion:

1. Ignition occurs within 10–50 ms. The rate of ignition decreases with increasing particle diameter.
2. Coals containing higher volatiles ignite faster even at relatively low temperatures. Biomass fuels having high volatile content can ignite quickly. For example, lignite coals and sub-bituminous fuels need not be pulverized to the same degree of fineness as low-volatile bituminous and anthracite coals, since they are rapidly ignited.
3. The volatiles are released in less than a duration of 5 ms.
4. While the volatiles are released, the particles are seen to swell about two to eight times their original volume.
5. In commercial furnaces, around 50% of a coal particle is observed to burn within 50 ms. Around 90–95% of the particle burns within 300 ms. The remaining 5% proceeds gradually toward complete burning. The type of fuel, temperature of the furnace, excess air and turbulence are the parameters affecting the combustion process.
6. Higher furnace temperature favors higher burning rates.
7. The time taken for complete burning depends on the square of the particle diameter.

8.   When the particle swells, its burning rate increases. However, the swelling causes the density to decrease and consequently it is carried away by the gas flow resulting in increased carbon loss through the fly ash.
9.   Turbulence and flow features such as recirculation and swirl seem to improve combustion efficiency.
10. Good combustion performance is obtained when ignition occurs around 75 mm to 300 mm from the nozzle exit. Ignition occurring too close to the nozzle may result in overheating of the nozzle.

Primary, secondary air flow rates, turbulent intensities and fuel type influence the point of ignition, rate of combustion and flame shape. The velocity of air and fuel is regulated through a value, which is slightly greater than the rate of ignition. If the exit velocity from the nozzle in the burner is too small, it usually causes a flashback. When the nozzle exit velocity increases beyond the ignition rate for the fuel type and the air–fuel ratio used, transient and unstable flame conditions are observed.

By proper adjustment of the primary and secondary air flow, either a long and penetrating flame or a short and intense turbulent flame is obtained in these burners. A high rate of heat release and high combustion efficiency is obtained from a turbulent flame. If multiple burners firing horizontally are placed opposed to each other, a high level of turbulence is produced in the zone of impingement. This causes an enhancement in the firing rate. Each horizontal burner is capable of firing around 400 kg of coal per hour. Therefore, in many installations, only a fewer burners are required.

Even though pulverized burners offer several advantages, there are a few significant disadvantages. Up to 85% of all the fine ash in the burner flows out through the boiler along with the flue gases. Owing to the high temperatures prevailing in the boiler, the ash softens and deposits on the furnace walls, boiler tubes and baffles. This is called a *slag* deposit. Around 50% of the fly ash is deposited in this manner. The slag deposit affects the heat transfer to the boiler, apart from blocking the flow in the narrow passages. These also cause operational issues and maintenance problems. Therefore, adequate equipment for periodic removal of slag is required to knock down the ash into the furnace, from where it can be removed. The amount of ash that passes through the boiler along with the flue gases may also be reduced by using fly-ash collectors.

Furnaces with pulverized fuel burners vary in design depending upon the way the ash is removed. In designs where the temperature is high enough and the ash melts, the furnace is built in a manner as to facilitate easy removal of the molten ash. The schematic of furnace where molten slag is continuously removed is shown in Fig. 6.10. A vertically fired pulverized burner is installed in this furnace as shown in Fig. 6.10. Part of the ash settles to the bottom of the furnace, where it mixes with a pool of molten ash. Since the flame passes over the pool of molten ash, ash remains in the molten state. The molten ash is gradually allowed to flow over a water-cooled slag container that opens into a water bath. The rapid quenching of the molten ash by the water bath converts it into a granular form. The granules are easily removed from the furnace.

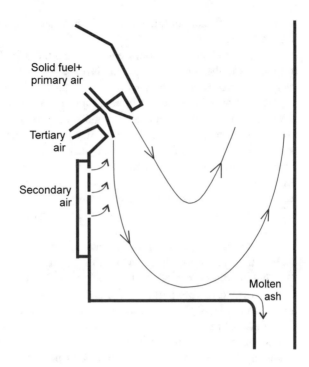

Fig. 6.10 Vertically fired pulverized fuel burner installed in furnace handling molten ash

Solid fuel+ primary air

Tertiary air

Secondary air

Molten ash

In a vertically fired furnace, the primary air and pulverized solid fuel are fed in downward direction directly into the combustion zone. Long burners with rectangular cross section are usually used, since a wide and thin flame sheet may be obtained from such burners. In vertical firing, the secondary air is generally supplied through small ports on the front wall placed downstream, where the hot gases flow downward. The secondary air supply can be controlled in order to make the flame shorter and intensified. A long diffusion flame may result when the secondary air is gradually supplied in the downward direction. A tertiary air supply is also provided in a co-flow manner to the primary air and fuel dust stream. This helps to regulate ignition and also enhances mixing. The primary and tertiary air flow rates are adjusted to regulate the extent of the flame inside the furnace so as to obtain a uniform temperature distribution within the furnace. Coals having low volatility, which require more residence time, and hence a longer flame, can be fired in this type of furnace. Since the flame length can be controlled by adjusting the secondary air supply, this method is equally good for firing high-volatile content fuels.

The design of a furnace in which the ash remains dry consists of a series of water tubes placed across the lower part of the furnace. These water tubes absorb the heat rapidly from the hot flue gases flowing through that zone. As a result, the space below the water tubes is relatively cool. Consequently, the ash, as it settles down in this space, also remains cool, and hence it does not melt and stays as a dry and dusty powder. Alternatively, the bottom wall may be inclined at an angle to the horizontal with water tubes installed beneath in order to maintain the wall at a lower temperature. The ash settling down on the cold wall remains dry and slides down the wall into the ash pit.

Figure 6.11 shows a pictorial view of a tangentially fired furnace. Here, the burners are located in the four corners of the furnace and each inject the jet of primary air and solid fuel dust in a direction that forms a tangent to a centrally located imaginary circle. Ignition of the fuel air stream injected by each burner is accomplished in the flame zone, which is located around the circle. An intensive turbulent field is created as the flames impinge. The hot flue gases fill the furnace and promote rapid and complete combustion as in a well-stirred reactor. The control of air and fuel flow rates is easier in these burners when compared to horizontally fired burners. High rates of firing with high combustion efficiency, even with little excess air is achievable in tangentially fired furnace. Figure 6.12 shows the carbon loss in the fly ash as a function of percentage excess air, in comparison between the different firing methods.

In order to avoid the primary disadvantages of the conventional pulverized burners such as high fuel crushing costs and loss of carbon in fly ash, cyclone burners are used with certain types of solid fuels. In these burners, as shown in Fig. 6.13, the coal is crushed to have particles less than 6 mm size and is fed along with primary air circumferentially into a compact cylindrical furnace typically having a diameter of around 2–3 m.

Secondary air, which is around 75–80% of the total air required, is also fed circumferentially into the chamber through another port. This ensures a thorough mixing

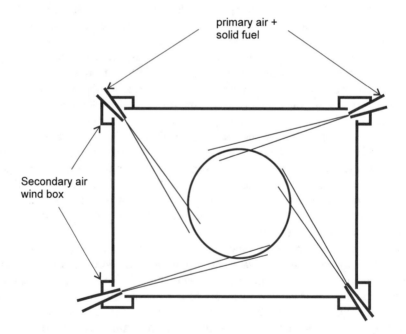

Fig. 6.11 Schematic of a tangentially fired furnace

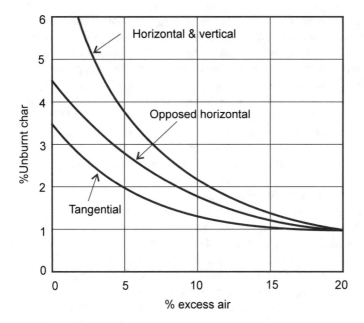

Fig. 6.12 Comparison of carbon loss in fly ash between different firing methods

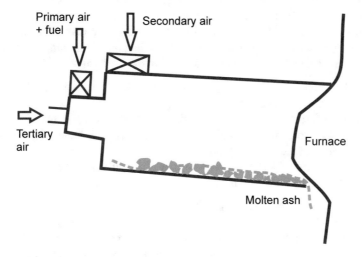

Fig. 6.13 Schematic of a cyclone furnace

of the fuel, air and hot products within the combustion chamber. The chamber is usually water cooled to avoid excessively high wall temperatures. The combustion products leave through a throat and flow into a furnace where water walls and other heat exchangers are installed. Approximately 5% of the total air required is supplied as tertiary air along the axial direction of the cylindrical chamber. This provides an axial velocity component to the gases and also ensures complete combustion of the particles suspended around the axis of the chamber. Temperatures close to 1700 °C are achieved by using around 5–15% excess air. At this temperature, the ash melts. Since the primary and secondary streams have circumferential velocities of around 100–150 m/s, the centrifugal force transports the molten ash toward the periphery of the cylinder. Intense heat release rates of more than 18,500 MJ/m^3 per hour are achievable in cyclone burners. The ash remains in the molten state and flows down through a slag tap hole into a water bath. To enable this, the cylindrical chamber is kept slightly inclined with respect to the horizontal. The cyclone principle involved here provides high heat release rates. This design allows only around 10% of the ash in the coal to enter the furnace. This results in low slag deposits in the furnace and hence fly ash removal systems are not necessary. The low excess air, high turbulence and high furnace temperatures lead to a high combustion efficiency. These furnaces can be operated with any grade of coal or other solid fuels. These burners may require the air to be preheated up to a temperature of around 450 °C.

The heat losses in the combustor should be estimated systematically, and the design and operating parameters are fixed in order to have minimum losses. Moisture and ash in the fuel cause uncontrollable losses in the heat and mass transfer processes, thereby adversely affecting the performance. High ash content coal is generally washed to reduce the ash content and dried to remove the moisture. As a part of clean coal technology, ash content is kept below 35%. The combustion efficiency decreases due to unburnt fuel species and carbon monoxide. Heat is lost due to removal of hot ash and hot dry flue gases. However, these losses are controllable. Combustion efficiency is improved by proper selection of the burner and furnace, based on the type of fuel used. A furnace should be designed to provide:

1.   mixing of fuel and oxidizer by using recirculation zones and turbulence,
2.   proper ignition temperatures,
3.   adequate furnace volume to give the gases enough residence time for the combustion to complete within the furnace, or enough grate area to complete the burning process on the grate,
4.   proper control of air and fuel supply at different firing conditions to ensure complete and efficient burning.

The heat carried away by the hot flue gases may be reduced by decreasing the temperature difference between the stack gases and the air entering the furnace and by limiting the mass of the flue gases produced per kg of fuel fired. Additional heat exchangers may be installed to the boiler and the furnace to extract the heat from and thereby cool the combustion gases leaving the boiler. A device called an economizer may be installed to absorb some of the heat from the flue gases and use it to heat the feed-water fed to the boiler. An air preheater may be installed to heat the primary air using the heat from the flue gases. The use of preheated air reduces unburnt gas emissions. Often the flue-gas temperature may be lowered by better operation of the existing installation and proper maintenance. For instance, the boiler tubes have to be frequently cleaned both internally and externally. This increases the rate of heat transfer and hence more heat from the flue gases is transferred to the boiler. The lowest temperature to which the flue gases may be cooled is limited in practice to a value significantly above the dew point temperature of the water vapor present in the flue gas mixture, in order to avoid condensation of water on the surfaces of boiler tubes and other walls. The mass of the flue gas produced per *kg* of fuel is reduced by using an optimum value of excess air and oxygen rich air to allow for complete combustion as well as acceptable furnace temperatures. In installations using grate burners with air-cooled walls, the air supply is typically between 150 and 160% of theoretical air. For installations with water-cooled walls, the air supply is around 140–150% of theoretical air. For pulverized burners, the air supply is in the range of 130–140% of excess air. In contrast, in pressure atomized liquid fuel burners, the air supply is around 125–130% of theoretical air and in gas fuel burners, it is in the range of 115–120% of theoretical air. Solid fuel burners consume a higher quantity of excess air. This is necessary in order to reduce two primary sources of atmospheric pollution from solid fuel fired furnaces: (1) smoke, resulting from unburnt carbon and volatiles and (2) ash and other dust particles, such as sand particles, which flow out the stack along with the combustion gases. It is quite possible to achieve smoke-free combustion of any solid fuel with a well designed and operated furnace.

The removal of fly ash and dust from the flue gases is more difficult and expensive than the elimination of soot or smoke. Any combustion equipment that burns coal will invariably emit some amount of fly ash, but the problem is especially acute with spreader stokers and pulverized burners. In the latter, as much as 85% of the ash particles escape through the stack along with flue gases. Dust and fly-ash collectors of different types are available to trap and collect the fly-ash and dust. All of these operate on some combination of the following strategies:

1. Imparting a change in the direction of the flue by properly placed baffles resulting in the separation of the heavier solid particles.
2. Imparting a centrifugal force from a vortex flow to separate the heavy dust.
3. Providing a sudden reduction of gas velocity, which allows the solid matter to settle.
4. Installing electrostatic precipitators to collect the ash on electrically charged plates.

The efficiency of an ash collector is estimated in terms of the mass of ash removed in one pass through the separator to the total amount sent through it. Electrostatic precipitators have efficiencies in the range of 90–95% and they are smaller than the mechanical separators. However, the cost of installation is higher.

Mitigation of emissions

Methane, CO_2 and nitrous oxide (N_2O) are greenhouse gases and are classified as global pollutants. Methane can be completely burnt using secondary air. The production of CO_2 is unavoidable while burning coal and biomass. Thus, techniques such as CO_2 sequestration is evolving to reduce the release CO_2 into the atmosphere. On the other hand, gases such as sulfur dioxide (SO_2), nitric oxide (NO) and volatile organic compounds contribute to regional pollutants. It is necessary to mitigate the release of these regional pollutants and nitrous oxide into the atmosphere. Techniques available to reduce the emissions of oxides of sulfur and nitrogen are discussed briefly in this section.

The sulfur content in the coal varies widely from 0.1 to 10%, based on the region from which it is obtained. Sulfur is present in one of three forms, namely pyritic, organic and sulfate. When coal burns, the sulfur is oxidized primarily to sulfur dioxide at temperatures between 800 and 900 °C, through the reaction,

$$S + O_2 \rightarrow SO_2.$$

A part of the SO_2 produced may react with excess oxygen to form sulfur trioxide, SO_3 through the reaction,

$$SO_2 + 0.5O_2 \rightarrow SO_3.$$

The formation of sulfur trioxide depends on gas residence time, temperature, pressure and excess air available in the furnace. The reaction is generally favored under high temperature and pressure conditions. Since the reaction is slow, only a small part of the sulfur dioxide converts into sulfur trioxide. The sulfur trioxide in turn reacts with the moisture in the flue gas and readily forms sulfuric acid, according to the reaction,

$$SO_3 + H_2O \rightarrow H_2SO_4.$$

The sulfuric acid condenses on cold surfaces. Limestone ($CaCO_3$) and dolomite ($CaCO_3.MgCO_3$) are two principal sorbents used for absorbing SO_2. The chemical reaction involving SO_2 with limestone that leads to its retention is,

$$CaCO_3 + SO_2 + 0.5O_2 \rightarrow CaSO_4 + CO_2.$$

This reaction does not take place in one step. The first step is called *calcination*, in which limestone decomposes to calcium oxide, CaO, and CO_2 through an endothermic reaction given as,

$$CaCO_3 \leftrightarrow CaO + CO_2.$$

The second step is called *sulfation*, in which CaO absorbs SO_2, forming calcium sulfate, $CaSO_4$, which is a relatively inert and stable solid that is handled easily. The overall reaction is expressed as,

$$CaO + SO_2 + 0.5O_2 \rightarrow CaSO_4.$$

Calcium oxide also reacts with SO_3 to form $CaSO_4$, when assisted by heavy metal salt catalysts. This reaction is expressed as,

$$CaO + SO_3 \rightarrow CaSO_4.$$

Dolomite, a compound of calcium carbonate and magnesium carbonate, is especially used in pressurized fluidized bed combustor because limestone does not calcinate at pressures above 3 *atmospheres*. The reactions involved are:

$$CaMg(CO_3)_2 \leftrightarrow CaCO_3 \cdot MgO + CO_2$$

$$CaCO_3 \cdot MgO + SO_2 + 0.5O_2 \rightarrow CaSO_4 \cdot MgO + CO_2$$

In this manner, based on the sulfur content in the solid fuel, proper amounts of limestone and dolomite are used to mitigate the release of sulfur oxides.

Sources for nitric oxides are the nitrogen present in the air and that present in the fuel itself. The fuel-bound nitrogen is oxidized to nitric oxide, NO, through a series of reactions. The volatile (air-bound) nitrogen appears as NH_3 or HCN. Ammonia (NH_3) may decompose into NO, while HCN results in the formation of N_2O. Around 77% of the fuel-bound nitrogen is oxidized to NO, and the rest appears as NH_3, which in turn is partly converted to nitrogen. Part of the NO formed may also convert back to nitrogen. The following methods are used to mitigate nitric oxides:

(1) Maintaining the furnace temperature in the range of 750–900 °C inhibits the oxidation of the nitrogen present in the air to nitric oxides. The amount of nitric oxides produced from fuel-bound nitrogen also decreases with a decrease in the furnace temperature.
(2) Lowering the excess air.
(3) Supplying the required air in stages rather than all at once. Such a staged air supply has a significant beneficial influence in reducing nitric oxides production, especially for solid fuels having high amounts of volatiles.
(4) Injection of ammonia (NH_3) in the upper section of the furnace or in the cyclone of a circulating fluidized bed boiler.

6.2 Gasification of Solid Fuels [16, 20]

From the discussions in previous sections, it is clear that the combustion of solid fuels such as coal and biomass is a relatively complex process. Control of air supply, fuel preparation and feeding, maintaining proper flow and temperature within the furnace, and removal of ash and other solid particles, are not straightforward. On the other hand, burning of liquid and gaseous fuels is relatively easy and more controllable. Liquid and gaseous fuels may be extracted from solid fuels using several techniques. Such fuels are called *synthetic* fuels. The synthetic gas or liquid fuels can be cleaned up and used efficiently in several applications.

The process of extracting gaseous fuels out of solid fuels is called the *gasification*. This has turned out to be a promising technique and is adopted in clean coal technology. Gasification of coal is different from carbonization of coal, in which coal is burnt in a retort in the absence of air and small amounts of gaseous fuels are extracted. However, gasification evolved from carbonization through a series of refinements to the latter. Over a period, the carbonization process was improved by supplying some amount of air gradually through a hot bed of coal so as to convert all the carbon in the coal primarily to carbon monoxide. Some amount of carbon dioxide which is produced is made to pass through the hot coal bed to produce more carbon monoxide. The reactions involved are surface reactions and these may be written as,

$$C + O_2 \rightarrow CO_2 \text{ (exothermic carbon oxidation reaction)}$$

$$C + CO_2 \rightarrow 2CO \text{ (endothermic reduction reaction called } Boudouard \text{ reaction)}$$

Collectively, these two reactions are exothermic. The gas mixture obtained through this process is termed as *producer* gas and it has a typical composition of 20–25% CO, 55–60% N_2, 2–8% CO_2 and 3–5% hydrocarbons. Owing to the significant amount of nitrogen that is present in producer gas, the heating value of the gas is low. In order to reduce the temperature of the bed, which may reach high values as a result of the carbon oxidation, steam is often supplied to facilitate the carbon-steam endothermic reaction,

$$C + H_2O \rightarrow CO + H_2 \text{ (endothermic carbon - steam reaction)}$$

The carbon-steam reaction yields *water–gas* (CO + H_2). Water–gas is used for synthesizing chemicals and is a good source of hydrogen. Treating the water–gas with steam oxidizes CO to CO_2 and increases the amount of hydrogen. This is called *water–gas shift* reaction, expressed as,

$$CO + H_2O \rightarrow CO_2 + H_2 \text{(slightly exothermic water - gas shift reaction)}$$

Hydrogen reacts with carbon, generally at elevated pressures, and in the presence of catalysts, to yield methane. This *carbon-hydrogenation* reaction is written as,

$$C + 2H_2 \rightarrow CH_4 \text{(exothermic carbon - hydrogenation reaction)}$$

In this manner, gasification process converts almost all the carbon in the solid fuels to gaseous fuels such as CO, H_2 and CH_4. The quantity of air used in this process is around 30–35% of the theoretical air required for the mass of solid fuel fed.

6.2.1 Types of Gasification Systems

Gasification systems are classified based on.

(1) the speed of operation or rate of burning,
(2) the direction of the supply of solid fuel particles and that of the gasification agents such as air/oxygen and steam, and
(3) the method of ash removal.

Based on the speed of operation, the gasification systems may be classified as (a) *fixed* or *moving bed*, (b) *fluidized bed* and (c) *entrained flow* gasification systems. The fixed bed system is the slowest and provides residence time of the order of a few hours to the particles. Fluidized bed gasification unit is faster than the fixed bed and provides residence time of the order of several minutes. The entrained flow gasification system is the fastest that provides residence time of the order of a few seconds.

Based on the direction of supply of the fuel, air/oxygen and steam, the gasification systems may be classified into (i) *updraft*, (ii) *downdraft*, (iii) *co-current* and (iv) *counter-current* gasification systems. In the updraft type, the air and steam are supplied through the bottom of the unit and the synthetic gas is removed from the top. In the downdraft gasification system, the gasification agent is supplied through the top and synthetic gas is extracted from the bottom. In these units, if the solid fuel and gasification agents move in the same direction, then it is called a co-current type. On the other hand, if the solid fuel and gases travel in opposite directions, it is termed as counter-current type. When air is used as the oxidizer, the system is called *air-blown* gasification system. It is called *oxygen*-blown when oxygen is used.

Based on the ash removal method, systems are classified as *dry-ash* type and *molten-ash (slag)* type. In the dry-ash type, the temperature of operation is such that the ash does not melt. On the other hand, in the slag-type systems, the operating temperature is high enough to melt the ash and the molten ash is removed through appropriate means. A summary of the classification is presented in Fig. 6.14.

6.2.2 Design, Performance and Emission Characteristics

Fixed bed gasification systems are in general counter-current systems. Here, the solid fuel is fed through the top and gasification agents are supplied through the bottom. The schematic representation of a fixed bed gasification system is shown in Fig. 6.15. This gasification process requires the least amount of oxidizer and steam and delivers synthetic gases with a higher heating value comparable to other systems. The operation cost is also lower than other gasification units.

However, it is mostly used for small-scale operations and scaling up to high power requirements is not possible. In this design, the superficial velocity of oxidizer and steam mixture is maintained below the minimum fluidization velocity of the solid

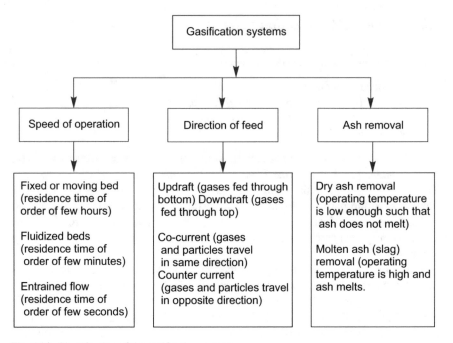

Fig. 6.14 Classification of the gasification systems

Fig. 6.15 Schematic of a
fixed bed gasification unit

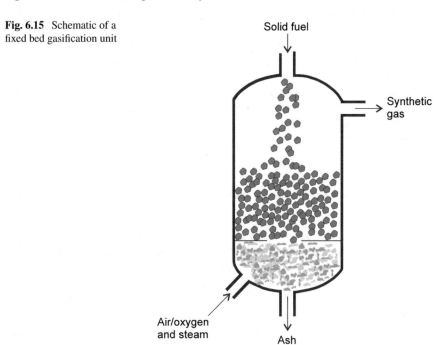

fuel particles. This causes the particles to be stationary and results in a non-uniform temperature distribution inside the gasification chamber. The chamber temperature varies from 430 to 1540 °C. This device is operated either in slag-ash (molten-ash) mode or in dry-ash mode. A slag-ash bed is usually operated with fine particles while the dry-ash bed is operated with coarse particles. For operation with a dry-ash bed, the chamber temperature should be between 430 and 1095 °C. The desirable particle size range for fixed bed units is 6–80 mm. The residence time of the coal is generally quite high, being of the order of hours.

The pictorial view of a fluidized bed gasification system is shown in Fig. 6.16. As in the case of fluidized bed combustors, here too, solid fuel particles having size less than 6 mm are used. Here, the superficial velocity of the oxidizer and steam mixture is kept higher than the minimum fluidization velocity of the solid particles and different regimes such as bubbling regime, turbulent regime and fast fluidization regime are realized. Gasification of the solid particles occurs in suspended mode. As a result, uniform mixing occurs and properties are more or less uniform within the chamber. The temperature difference between the gas and the solid phases is quite small. These systems are also counter-current type in general. The synthetic gas produced after the gasification process leaves the chamber through the top. The residence time of fuel particles inside the chamber is of the order of several minutes. Usually, these units operate below the ash melting temperature to avoid ash agglomeration and clinker formation. The operating temperature is in the range of 870–1038 °C. Depending

Fig. 6.16 Schematic of a fluidized bed gasification unit

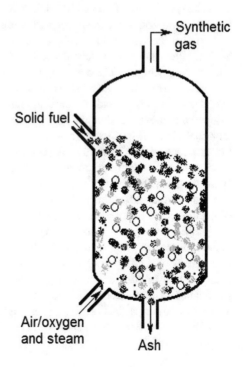

Fig. 6.17 Schematic of an entrained flow gasification unit

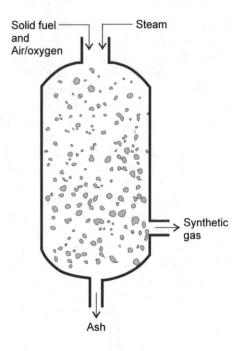

upon the type of solid fuel employed, it is possible to realize higher efficiencies due to uniform mixing of the feedstock and oxidizer.

Entrained bed gasification units are generally operated in co-current and downdraft mode as shown schematically in Fig. 6.17. The synthetic gas yield from these units is usually higher. The operating temperature ranges between 930 and 1650 °C. The preferred average particle size is less than 0.15 mm. Thorough mixing of the fuel particles and oxidizer leads to uniform chamber temperature and higher efficiency compared to moving and bubbling bed systems, especially when operated with high rank coals. In these gasification systems, a variety of feedstock can be used. However, low ash content, less than 20%, is preferred. The main disadvantages of this gasification system is the requirement of higher amounts of oxidizer and production of synthetic gas with a relatively lower calorific value.

Factors which affect the gasification process are the ash content in the solid fuel, characteristics of the volatiles, reactivity of the char and the calorific value of the fuel. The gasification system is chosen based on these factors. Based on the gasification system chosen, proper operating conditions have to be maintained. The design should take care of variability in the fuel feed stock. This includes a range of operating pressure, temperature, air-to-coal ratio and steam-to-coal ratio. In general, all the gasification systems need auxiliaries such as air preheater, steam generator, cyclone to remove larger particles and gas cleaning system that has dust and tar removal capabilities. The cleaned gas is stored carefully in sealed containers at the required pressure.

Many solid fuel burning and gasification systems have been developed for use in a variety of applications. Coal, wood, biomass and even solid wastes are being burned and gasified. In some of the designs, the solid fuel, especially biomass, is gasified and subsequently the synthetic gas produced is immediately burned. Such burners are used in domestic cooking and heating applications.

Review Questions

1. Classify the gasification and burning systems used for solid fuels.
2. Compare the merits and demerits of different overfeed systems.
3. What is the use of arch in travelling grate stoker?
4. What is the importance of fuel flexibility in coal combustors?
5. Explain the working process of fluidized bed combustors.
6. List the basis on which pulverized coal burners are designed.
7. What are the heterogeneous reactions involved in gasification of solid fuels?
8. What is water gas shift reaction? Explain its usefulness in the gasification process.
9. Briefly discuss the merits and demerits of different gasification systems.
10. What are the factors affecting the burning and gasification process of solid fuels?

Exercise Problems

1. A solid fuel has the following constituents on mass basis: 65% carbon, 6% H_2, 10% O_2, 4% N_2 and 15% ash. When it is burnt in a furnace, the carbon content in the solid residue left over is found to be 25%. Volumetric analysis of the flue gas gave the following results: 14% CO_2, 1% CO, 3.5% O_2 and 81.5% N_2. Find the actual air supplied, percent excess air and mass fractions of products.
2. The devolatilized gases from a coal is assumed to be a compound of the form $C_xH_yO_zN_w$. Values of x, y, z and w can be obtained from proximate and ultimate analyses. The coal has 60% fixed carbon, 40% volatile matter based on dry ash-free basis. On the same basis, the ultimate analysis is given as C—0.75, H—0.05, O—0.16 and N—0.04. Find the stoichiometric air-to-fuel ratio to burn the volatiles.
3. A grate burner is fed with anthracite coal (95% C by mass) particles of mean size 25 mm (equivalent sphere diameter) at a temperature of 25 °C. Its calorific value is 27 MJ/kg, specific heat is 1.26 kJ/kg-K and its specific gravity can be taken as 1.42. The flow of gases is such that the average heat transfer coefficient is 15 W/m²-K. When the particle reaches 900 °C, the average mass loss rate of carbon is 2×10^{-9} kg/s. Determine the approximate width and length of the moving grate to achieve a power rating of 200 kW assuming 90% combustion efficiency.
4. A fluidized bed combustor operates at 5 bar pressure and 900 °C bed temperature. Coal particles with a mean (sphere) diameter of 6 mm and average density of

1250 kg/m^3 are used. Calculate the minimum air velocity from the distributor plate that will fluidize the coal particles, by just equating the drag force to the particle mass. The coefficient of drag may be assumed as 0.45.

5.   In a coal firing furnace of 1500 kW power rating, a coal with a calorific value of 23 MJ/kg is fed. The sulfur content in the coal is 2.7% by mass. Crushed limestone is added to mitigate the release of sulfur oxide to the atmosphere. Limestone contains 88% of $CaCO_3$. If the combustion efficiency is 80%, determine the feed rate of limestone.

Chapter 7
Alternative Fuels

Fossil fuels such as coal and petroleum products are depleting at a rapid rate due to the increasing consumption and decreasing availability. Therefore, alternative fuels, which can be a substitute to the fossil fuels, are becoming important. Alternative fuels can either be used as neatly they are or can be blended with the fossil fuels and used in many applications. Several alternative fuels have been used in the past when fossil fuels were not available as much required at those times. However, these fuels have been discontinued over a period of time either due to increased availability of the fossil fuels or due to the non-economical nature of those alternative fuels. For example, people were running their engines with vegetable oils obtained straight from plant seeds, when fossil diesel was not available in plenty. When the petroleum production was increased, the usage of vegetable oils became non-economical when compared to the usage of petroleum diesel. Currently, several alternative fuels are revived and many more new alternative fuels are being invented by researchers, keeping in mind the depleting fossil fuels and the emissions resulting from their usage. In Chap. 1, an introduction to several fossil and alternative fuels was given. In this chapter, brief discussions of the methods of producing renewable alternative fuels, their characterization, performance in typical applications and related emissions are provided.

7.1 Production and Characterization of Alternative Fuels

Alternative fuels may be classified as *renewable* and *non-renewable* fuels. Renewable fuels are obtained from naturally available or naturally grown raw material called *biomass.* The term biomass represents vegetable crops, waste from vegetation and crops, forest products, animal wastes and algae. Biomass such as wood, wood chips, husks and dried leaves can be directly fired in small stoves or boilers. They require a little fuel preparation such as sun-drying and reduction of size. They have been used in cooking and heating applications extensively. On the other hand, biomass fuels such as vegetable seeds, animal wastes and algae need considerable preprocessing before they can be used as fuels in practical applications. Depending upon the source and the type of the end product obtained, three methods are available for producing biomass-based alternative fuels. They are *bio-based*, *agro-based* and *thermal-based* methods.

Bacterial fermentation of sugar to ethanol is an example for a bio-based method. Ethanol is produced from fermentation of pre-treated starch and cellulose biomass. The production of biogas from animal wastes such as cow dung or other types of organic wastes also follows the bio-based method.

Production of biodiesel from vegetable oils extracted from vegetable seeds grown for this purpose, or from the oil extracted from algae, follows the agro-based method. In general, food crops such as corn, sugarcane, coconut and other edible vegetable seeds are not preferred as alternative fuel sources. Non-edible vegetable seeds such as neem, jatropha and karanja have to be grown to meet the fuel needs.

Gasification and liquefaction of several biomass fuels to produce different types of gaseous and liquid alternative fuels involve thermal-based methods. Pyrolysis and gasification involve heating of the biomass fuel to a certain temperature range in a controlled environment so as to generate gaseous fuels. Fischer-Tropsch (F-T) process is used to extract liquid hydrocarbon fuels from biomass. Methods such as steam reformation and supercritical steam gasification are used to produce hydrogen from different types of gas mixtures extracted from biomass. Research has been going on to derive energy from municipal solid wastes and segregated plastics as well.

Figure 7.1 provides a summary of different methods used for producing alternative fuels from different types of biomass.

Biogas produced from biological fermentation of food and animal wastes contains 60–80% of methane and 20–40% of carbon dioxide by volume. Its calorific value lies between 30,000 and 32,000 kJ/kg. Biogas is extensively used in cooking and heating applications, especially in rural areas, where it can be produced in required quantities using a simple bio-digester.

Ethanol produced from biomass is quite pure and has been successfully employed for automotive applications after blending it with gasoline. At present, a 20–30% blend of ethanol with gasoline is used worldwide in automobiles. Although the emissions from burning the blend are better, the energy economics of ethanol blended gasoline is not favorable.

Biodiesel is a processed or transesterified vegetable oil. Methanol or ethanol is used in the transesterification process, where the important process is the removal of glycerin in the vegetable oil. After transesterification, the viscosity of the biodiesel reduces to a value conducive for use in atomizers designed for diesel fuel. Biodiesel is a mixture of mono-alkyl esters of long-chain fatty acids. It has a lower calorific value than diesel. However, it burns much cleaner owing to the presence of oxygen

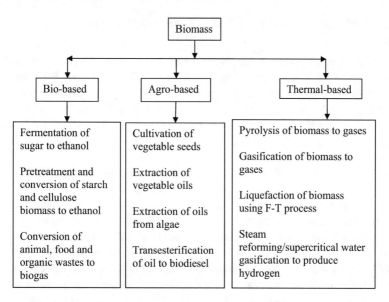

Fig. 7.1 Different methods for producing alternative fuels from biomass

atoms. At present in unmodified diesel engines, a blend of 20% biodiesel and 80% diesel can be used to reduce the emissions.

Biomass, when heated to a temperature of 400–600 °C, undergoes pyrolysis and produces a mixture of gases (producer gas). On the other hand, gasification, as discussed in Chap. 6, is a process carried out with limited oxygen supply, at a higher temperature between 850 and 1100 °C. This process produces a mixture of gases called the synthetic gas. To increase the yield of hydrogen, steam gasification or steam reforming is often used. Steam reacts with CO and methane to produce hydrogen. Nickel-based catalysts are used to improve the yield of hydrogen. Super-critical steam gasification of biomass is used to produce hydrogen extensively. Even wet biomass can be used in this process. By controlling the steam to biomass ratio and the supply temperature of the steam, the hydrogen yield can be increased even to around 50% by volume of the mixture. Fischer–Tropsch process produces liquid hydrocarbons from a gas mixture predominantly containing hydrogen and carbon monoxide, obtained from biomass gasification. A process called hydro-cracking is used to produce diesel-type liquid fuel from the heavy products of F-T process. Distillation process is used to separate out the lighter liquid fuels as by-products.

7.2 Performance and Emission Characteristics

Biogas has a low calorific value due to high CO_2 content. Therefore, its specific fuel consumption is quite high. Also, the burning characteristics and flame structure are affected by the CO_2 content. Proper primary and secondary aeration is required to efficiently burn the biogas in small-scale applications. Biogas is also used as an additional fuel in IC engines. Based on the organic waste from which the biogas is produced, some amounts of impurities may be present and hence the gas should be cleaned before use.

Using ethanol obtained from biomass along with gasoline in transportation applications has become common and also mandatory in many countries. Addition of ethanol is small quantities has shown improved performance and reduced emission of CO, volatile organic compounds, particulate matters and so on. However, the emission characteristics of the E85 blend, which contains 85% ethanol, commonly used in the USA and Brazil, have not been consistent—increased emissions have been reported in a few cases. Usage of cellulose biomass to produce bioethanol instead of corn or sugar based sources, the so-called grassoline, is also being pursued now.

Biodiesels blended with diesel have been able to reduce the emissions of CO, unburned hydrocarbon and particulate matter. Studies using 100% biodiesel in engines indicate that these pollutants reduce by around 50–60% when compared to that produced by the fossil diesel. However, the emission of nitric oxides has been observed to increase by around 10%. Additionally, issues in using large amounts of biodiesel in engine are the formation of engine deposits, aging (changes in the composition, thermo-physical and calorific properties) and corrosion. Therefore, the composition of the biodiesel–diesel blend should be carefully chosen in order to realize the best characteristics of biodiesel.

Producer and synthetic gases, obtained from direct pyrolysis or gasification, contain a large volumetric percentage of nitrogen. This brings down the calorific value. Nitric oxide formation is inevitable when these gases are fired. This can be controlled by ensuring proper temperature distribution and secondary aeration. Further, these gases have to be cleaned to remove solid particles of ash and unburned char, and tar. The tar, if not removed, can condense in the fuel pipes, resulting in clogging. If the biomass source has sulfur, then sulfur oxides are also formed. During gasification process, limestone is customarily used to avoid sulfur oxide formation.

Liquid fuels obtained from the F-T process are further processed to obtain diesel- or gasoline-type fuels, which can be employed in IC engines. Heavier liquid fuels can be used in industrial furnaces. These liquid fuels have properties much closer to fossil petroleum fuels and burning of these fuels produce emissions similar to the fossil fuels.

7.3 Future Scope

It was mentioned in Chap. 1 that the issues associated with the sustainability of a biofuel, such as growing the required amount of the vegetation for its production, have not been resolved fully. Further, issues related to faster conversion techniques, scaling up of the production processes and maintaining the supply chain, have to be addressed systematically.

First of all, it is essential to study the relevant processes involved in the production of various biomass-based fuels, depending upon the local feedstock availability around a proposed refinery. Then the processes have to be optimized for the type of feedstock available and the products that can be produced. Technical and financial feasibilities have to be studied thoroughly.

Specifically, technologies for the production of hydrocarbon fuels from biomass have to be studied in detail. The synthetic and metabolic pathways to produce such fuels are currently not well established. The complete physics involved in the fermentation processes for bio-based conversion also needs to be clearly understood. The additional processes required for fuel recovery, efficient conversion and purification are to be streamlined. Lifecycle analysis of the processes to produce these fuels has to be carried out in order to understand the sustainability of the fuel source.

Studies are required to develop biomethane using feedstock such as cellulosic materials and organic wastes. Focused research on the proper methods to pre-treat the cellulose biomass for the efficient production of bioethanol, is required. This includes the choice of proper enzymes for hydrolysis of cellulose. Oil from pyrolysis products is produced by thermal decomposition of biomass fuels in the absence of oxygen. During this process, apart from the oil, char and other gases are also produced. This oil could be a potential replacement for furnace oil. Proper choices of the feedstock, optimization of the processes to get maximum oil yield and cost-effective technology have to be established through systematic research.

Finally, butanol derived from biomass is a potential biofuel, classified as an advanced biofuel, because of its higher energy content than ethanol. Biobutanol blended with gasoline can be used in gasoline engines. Feedstock such as sugarcane bagasse, wheat and rice straw and wastes from paper industries can be used in the production of biobutanol. Proper methodologies for pre-treatment and hydrolysis, fermentation and recovery of alcohols have to be developed. The processes should also be made cost effective.

In summary, research on the production and utilization of alternative biofuels are very much required to complement and substitute the fossil fuels in many applications. However, systematic research and analyses are required to address the sustainability, technical and economical issues associated with the renewable biofuels.

Review Questions

1. How are alternative fuels classified?
2. List the methods available for producing renewable fuels from biomass.

3.   What is the typical composition of biogas?
4.   What are the advantages of biodiesel over fossil diesel?
5.   List the differences between pyrolysis and gasification processes.
6.   What are the benefits of producing liquid hydrocarbon fuels from biomass?
7.   Write about the limitations of using ethanol in IC engines.
8.   What are the typical pollutants produced during the combustion of synthetic gas?
9.   What are the challenges involved in the production of hydrocarbon fuels from biomass?
10. List the advantages of biobutanol as an alternative fuel.

Chapter 8
Numerical Modeling of Laminar Flames

In order to understand the characteristics of a flame, sufficient details of its flow, temperature and species fields are required. Experimental measurements of these quantities in flames are difficult to carry out. In several scenarios, usage of intrusive or probing instruments, such as Pitot tubes, thermocouples, microprobes and pressure transducers, may not be possible due to accessibility issues. Further, intrusive measurements may also affect the flame structure and stability. The non-intrusive techniques such as laser-based diagnostics are much costlier and relatively tougher to use in practical furnaces. Therefore, mathematical modeling is a viable option to predict the features of a flame, which is a chemically reacting flow phenomenon. Due to the nonlinearities and coupled nature of the governing equations involved in a reacting flow, various numerical parameters have to be employed in solving them. Due to the invention of sophisticated computational resources at much lower costs, when compared to laser-based diagnostics, numerical modeling has become an economical as well as a comprehensive tool to simulate flames from burners in furnaces and stoves. In this chapter, details of numerical modeling of laminar flames, which are quite fundamental in nature, are presented systematically.

8.1 Governing Equations

A reacting flow is constituted by a mixture containing multiple species. The concentration (or the mass fraction) of each species and its temperature vary at different locations. As a result of this, the properties of individual species, as well as those of the mixture, exhibit spatial variations. The species move through the mixture by convection (bulk flow) as well as by diffusion. Concentration gradient drives the species from higher concentration to lower concentration locations through a phenomenon called ordinary or molecular diffusion. Fick's law, considering ordinary diffusion, is used to evaluate the total mass flux of ith species, and is written as,

© The Author(s), under exclusive license to Springer Nature Switzerland AG 2022   177
V. Raghavan, *Combustion Technology*,
https://doi.org/10.1007/978-3-030-74621-6_8

$$\rho_i \vec{v}_i = \rho Y_i \vec{V} + \rho_i \vec{v}_{i,\text{diff}} \tag{8.1}$$

Here, $\vec{v}_i$ is the velocity vector of ith species, $\vec{V}$ is the velocity vector of the mixture, ρ is the mixture density ($= \sum \rho_i$), ρ_i is the density of ith species, and $\vec{v}_{i,\text{diff}}$ is the diffusion velocity vector of the mixture. The total flux of ith species is calculated as the sum of its convective and diffusion fluxes. Equation (8.1) is also written as,

$$\vec{v}_i = \vec{V} + \vec{v}_{i,\text{diff}} \tag{8.2}$$

That is, the species velocity is calculated as a sum of mixture (or bulk) velocity and diffusion velocity. The species diffusion also occurs due to the notable gradients in temperature, pressure and body forces, apart from concentration gradients (ordinary diffusion). The diffusion velocities of all the species in the mixture are calculated using Stefan–Maxwell equations, which include ordinary diffusion, thermal diffusion (Soret effect), pressure diffusion and diffusion by body forces. They are given as,

$$\vec{\nabla} X_i = \sum_{j=1}^{N} \left(\frac{X_i X_j}{D_{ij}} \right) \left(\vec{v}_{j,\text{diff}} - \vec{v}_{i,\text{diff}} \right) + (Y_i - X_i) \left(\frac{\vec{\nabla} p}{p} \right) + \frac{\rho}{p} \sum_{j=1}^{N} Y_i Y_j \left(\vec{f}_i - \vec{f}_j \right)$$
$$+ \sum_{j=1}^{N} \left(\frac{X_i X_j}{\rho D_{ij}} \right) \left(\frac{D_{T,j}}{Y_j} - \frac{D_{T,i}}{Y_i} \right) \left(\frac{\vec{\nabla} T}{T} \right) \tag{8.3}$$

In Eq. (8.3), X is mole fraction, Y is mass fraction, D_{ij} is molecular binary diffusivity, which governs molecular diffusion between any two species i and j, D_{Ti} is thermal diffusion coefficient and f_i is the body force vector of ith species. Molecular binary diffusivity, D_{ij}, is a property associated with a pair of species, i and j. It can be calculated as a function of pressure, temperature and molecular masses of species i and j. Further, the N diffusion velocities have the constraint,

$$\sum_{i=1}^{N} Y_i \vec{v}_{i,\text{diff}} = 0$$

Similarly, the sum of thermal diffusion coefficients of all species is zero. By solving N equations with the above constraint, the diffusion velocities of all species are obtained. The diffusion velocities calculated from Eq. (8.3) include ordinary, thermal, pressure and body force-induced diffusion processes. The solution of Eq. (8.3) is computationally intensive.

In general, for low-speed reacting flows, diffusion velocities from pressure gradients and body forces are negligible. From a simplified method, the ordinary diffusion velocity is calculated as,

$$\vec{v}_{i,\text{diff},X} = -\frac{D_{im}}{Y_i} \vec{\nabla} Y_i \tag{8.4}$$

Here, Y_i is the mass fraction of ith species ($= \rho_i/\rho$) and D_{im} is the diffusion coefficient (in m^2/s). The value of diffusion coefficient dictates the rate at which the ith species diffuses directly into the mixture. The estimation of D_{im} relies on a mixing rule, written in terms of mole fraction of species (X) and binary diffusivity (D_{ij}). It is given as,

$$D_{im} = \frac{(1 - X_i)}{\sum_{j \neq i} \left(\frac{X_j}{D_{ij}} \right)} \tag{8.5}$$

However, a correction is applied to the diffusion velocity calculated using Eqs. (8.4) and (8.5), to account for nonzero Lewis number (α/D_{im}), where α is thermal diffusivity. The diffusion velocity correction is obtained by iteratively solving equations written in terms of mass fractions of species, their molecular weights, binary diffusivity of pair of species $i - j$ and effective diffusion coefficient of ith species, D_{im}. Then, the correction is added to Eq. (8.4). It is expressed as,

$$\delta \vec{v}_{i,\text{diff},X} \sum_{j \neq 1} \frac{Y_j}{\text{MW}_j D_{ij}} - \sum_{j \neq 1} \frac{Y_j}{\text{MW}_j D_{ij}} \delta \vec{v}_{j,\text{diff},X} = \sum_{j \neq i} \frac{1}{\text{MW}_j} \left(1 - \frac{D_{jm}}{D_{ij}} \right) \nabla Y_j$$

Out of N equations given above, one is replaced by the following, which is used as a constraint.

$$\sum_{i=1}^{N} Y_i \delta \vec{v}_{i,\text{diff},X} = \sum_{i=1}^{N} D_{im} \nabla Y_i$$

The thermal diffusion (Soret) velocity is calculated as,

$$\vec{v}_{i,\text{diff},T} = -\frac{D_{T,i}}{\rho Y_i} \frac{\vec{\nabla} T}{T} \tag{8.6}$$

The thermal diffusion coefficient has units of kg/ms and is calculated using empirical, composition-dependent correlation. The diffusion velocity due to ordinary and thermal diffusion is obtained by adding Eqs. (8.4) and (8.6). This procedure is computationally less intensive. Any velocity measuring probe measures the mixture velocity. Similarly, solution to mass and momentum balance equations using a numerical model reveals the mixture velocity. The diffusion velocity is calculated using Eq. (8.3) in a comprehensive manner or using Eqs. (8.4) and (8.6), in a simplified manner. However, there is no direct way to measure or solve for the species velocity. It is calculated using Eq. (8.2).

The conservation equations are written for a control volume, which is bounded by the control surfaces, through which the multi-component mixture flows in and out. The fluid is Newtonian in nature and obeys continuum principle. The conservation equations are partial differential equations, consisting of four terms: time-dependent term that represents accumulation of a quantity within the control volume, convective terms, which are nonlinear terms, diffusion terms, which are linear second-order derivative terms and source terms, which may be nonlinear. Conservation of each species (say species A) in the multi-component mixture is governed by,

$$\frac{\partial}{\partial t}(\rho Y_A) + \nabla.\left[\rho Y_A\left(\overrightarrow{V} + \overrightarrow{v}_{A,\text{diff}}\right)\right] = \dot{\omega}_A^m \tag{8.7}$$

The rate of mass of species A accumulated in the control volume plus the net mass efflux (outgoing flux—incoming flux) of species A through the surfaces of the control volume is equal to the net rate of its production (or destruction), denoted by the term in the right-hand side ($\dot{\omega}_A^m$), which has units of kg/m^3 s. The net mass efflux of species A includes the fluxes due to the convection as well as diffusion. In a multi-component mixture with N components, ($N - 1$) species conservation equations are solved to obtain their mass fractions. The mass fraction of the Nth species is obtained using the identity, $\sum Y_i = 1$. The Nth species is usually kept as an inert species such as nitrogen.

If conservation equations of all the species are summed up, the conservation of the mass of the mixture is obtained. The net reaction rates of reactant species (consumption rates) cancel out the net reaction rates of product species (production rates), making the right-hand side of Eq. (8.7) to go to zero. Similarly, the sum of diffusion fluxes of all the species constituting the mixture is also zero. Subsequently, the conservation of mass of the mixture is written as,

$$\frac{\partial}{\partial t}(\rho) + \nabla.\left(\rho \overrightarrow{V}\right) = 0 \tag{8.8}$$

The mass of the multi-component mixture accumulated within the control volume plus the net efflux of the mixture mass through the surfaces of the control volume is equal to zero. For reacting flows observed in burners and stoves, the velocities involved are quite low and the flow is incompressible in nature. In several applications involving flow processes, the total pressure remains almost a constant. In these scenarios, the density of the mixture varies due to variations in temperature and composition of the mixture. It behaves as a fluid property and is evaluated as other properties, which vary with temperature and concentration of species. In an incompressible flow, the continuity equation is used to couple velocity and pressure, that is present in momentum equations. Velocity correction and pressure correction terms are arrived at in terms of continuity residue (nonzero) value.

Momentum conservation follows Newton's second law of motion. For fluids, it is stated as, the rate of change of momentum in a control volume is equal to sum of all forces acting on the control volume. It is written as,

$$\frac{\partial\left(\rho \overrightarrow{V}\right)}{\partial t} + \nabla.\left(\rho \overrightarrow{V} \overrightarrow{V}\right) = \nabla : \tau_{ij} - \nabla p + \overrightarrow{B} \tag{8.9}$$

In Eq. (8.9), first term in the left-hand side represents the rate of change of momentum per unit volume, the second term in the left-hand side is the convective term, $\nabla : \tau_{ij}$ is the gradient of the stress tensor (τ_{ij}), which includes normal and shear stresses at the control surfaces, ∇p is the pressure gradient and the last term in the right-hand side, $\vec{B}$ is the body force vector, which is evaluated as,

$$\vec{B} = \rho \sum_{k=1}^{N} Y_k \vec{f}_k \qquad (8.10)$$

Here, $\vec{f}_k$ is the volumetric force per unit mass acting on k^{th} species. This term can be neglected in several scenarios. The stress tensor can be written in terms of velocity gradients. In tensor form, it is written as,

$$\tau_{ij} = \left(\mu' - \frac{2}{3} \mu \right) \frac{\partial u_k}{\partial x_k} \delta_{ij} + \mu \left(\frac{\partial u_i}{\partial x_j} + \frac{\partial u_j}{\partial x_i} \right) \qquad (8.11)$$

Here, μ is the dynamic viscosity of the fluid mixture and μ' is the secondary (or bulk) viscosity. For incompressible fluid flow, the first term in the right-hand side of Eq. (8.11) is usually negligible.

Energy conservation is stated as *the rate of accumulation of total energy* plus *net efflux of total energy* is equal to *net rate of energy addition from surroundings* plus *rate of heat added by sources* plus *net rate of work done on the system by surroundings*. Here, the total energy per unit volume, neglecting the potential energy component, is the sum of internal energy (ρu) and kinetic energy $\rho \left(\vec{V} \cdot \vec{V} \right)/2$, per unit volume. It is written, per kg of the mixture, as,

$$e = u + \frac{v^2}{2} = \sum Y_i u_i + \frac{v^2}{2} = \sum Y_i h_i - p/\rho + \frac{v^2}{2}$$

Energy fluxes coming in and leaving out of the control volume are due to flow of species through the control surfaces, heat diffusion due to temperature gradient (Fourier conduction) and that due to concentration gradients at the control surfaces. Energy flux due to flow of species is written as a sum of enthalpy flux of the mixture due to convection, sum of enthalpy fluxes of all species due to their diffusion and convective fluxes of pressure and kinetic energies, given by,

$$\rho \vec{V} h + \rho \sum Y_i h_i \vec{v}_{i,\text{diff}} + \rho \vec{V} \left(-\frac{p}{\rho} + \frac{v^2}{2} \right) \qquad (8.12)$$

Here, h is the specific enthalpy of the mixture. It may be noted that the first and third terms in Eq. (8.12) are convective terms and the second one is a diffusion term. The heat flux due to conduction is written in terms of the thermal conductivity of the mixture and temperature gradient, as, $-\lambda \nabla T$. The energy flux due to concentration gradients is the reverse (secondary) effect of thermal diffusion (Soret effect), and it is called Dufour effect. It is written as,

$$R_u T \sum_i \sum_j \left(\frac{X_j D_{T,i}}{MW_i D_{ij}} \right) \left(\vec{v}_{i,\text{diff}} - \vec{v}_{j,\text{diff}} \right)$$

Here, MW_i is the molecular mass of ith species and R_u is the universal gas constant. Dufour effect is also not significant in several scenarios. The net diffusive flux of energy is written as,

$$\vec{q} = -\lambda \nabla T + \rho \sum_{i=1}^{N} Y_i h_i \vec{v}_{i,\text{diff}} + R_u T \sum_i \sum_j \left(\frac{X_j D_{T,i}}{MW_i D_{ij}} \right) \left(\vec{v}_{i,\text{diff}} - v_{j,\text{diff}} \right) + \vec{q}_R$$

(8.13)

The last term in the right-hand side of Eq. (8.13) is the flux due to radiative heat transfer. It may be noted that in many occasions, radiation losses are handled using a source term. In such cases, the last term representing radiative flux is not included in Eq. (8.13).

If the external force per unit mass of kth species is represented by $\vec{f}_k$, then the work interaction with the control volume due to body force of the species is written as,

$$\rho \sum_{k=1}^{N} Y_k \vec{f}_k \left(\vec{V} + \vec{v}_{k,\text{diff}} \right)$$

The viscous and the pressure forces also cause work interaction with the control volume. These arise due to the last term in Eq. (8.12), which is expressed using an equation that results from taking dot product of velocity vector and momentum equation. The resultant terms are written using substantial derivative (D/Dt) for pressure and tensor product of velocity vector and velocity gradient. This is expressed as,

$$\frac{Dp}{Dt} + \tau_{ij} : \nabla . \vec{V}$$

Here, the second term is called the viscous dissipation. The substantial derivative is written as,

$$\frac{D}{Dt} = \frac{\partial}{\partial t} + \vec{V} . \nabla$$

The above work interactions are often negligible for low-speed incompressible reactive flows. For incompressible laminar flows, the energy conservation equation, including convective, diffusive, radiation fluxes and source term ($\dot{Q}$), and excluding

Dufour and work interaction terms, is written as,

$$\frac{\partial}{\partial t}(\rho h) + \nabla \cdot \left(\rho \vec{V} h\right) = -\nabla \cdot \left(-\lambda \nabla T + \rho \sum_{i=1}^{N} Y_i h_i \vec{v}_{i,\text{diff}} + \vec{q}_R\right) + \dot{Q} \quad (8.14)$$

The mixture enthalpy, h, is calculated using the standard specific enthalpy of each species, h_i, and its mass fraction, Y_i, as, $h = \sum Y_i h_i$. The standard enthalpy (in J/kg) is written in terms of enthalpy of formation and the sensible enthalpy as,

$$h_i(T) = h_{f,i}^o(T_{\text{ref}}) + \Delta h_i(T) \quad (8.15)$$

The sensible enthalpy is calculated by integrating the temperature-dependent specific heat at constant pressure of the species between the temperature at the reference state (T_0) and the temperature of the given point. This is written as,

$$\Delta h_i(T) = \int_{T_0}^{T} c_{pi} \, \mathrm{d}T$$

The energy conservation is written in terms of temperature as,

$$\rho c_p \frac{\partial}{\partial t}(T) + \rho c_p \vec{V} \cdot \nabla T = \nabla \cdot (\lambda \nabla T) - \sum \dot{\omega}_i^m h_i$$
$$- \rho \sum_{i=1}^{N} Y_i \vec{v}_{i,\text{diff}} c_{pi} \nabla T - \nabla \cdot \vec{q}_R + \dot{Q} \quad (8.16)$$

The derivation of Eq. (8.16) uses the definition, $dh = c_p dT + \sum h_i \nabla Y_i$, and species conservation equation multiplied with h_i and summed up for all species. Specific heat of individual species (c_{pi}), standard enthalpies of all species (h_i), calculated as a function of temperature, and specific heat of the mixture (c_p), calculated as a function of concentrations of the species, are required in solving the energy conservation.

8.1.1 Calculation of Thermo-physical Properties

This section reports the methodology to evaluate properties used in the governing equations. In laminar flames, flow is incompressible. Unless the operating pressure is much high, as a result of high temperature, the density is low. Therefore, the ideal gas equation of state is employed when the pressure is much less than the critical pressure of the mixture. Density is a fluid property, and its value for the fluid mixture is evaluated using ideal gas equation of state. This is written as,

$$\rho = p/R_{\mathrm{mix}} T$$

Here, R_{mix} is the specific gas constant of the mixture, calculated using the universal gas constant and molecular weight of the mixture, as, R_u/MW_{mix}. The molecular weight of the mixture is in turn calculated using the chain rule using mass fraction (Y_i) or mole fraction (X_i), written as,

$$MW_{\mathrm{mix}} = \sum X_i MW_i = 1/ \sum \frac{Y_i}{MW_i}$$

For any pure component, its critical temperature (T_c), critical pressure (p_c), critical volume (V_c), acentric factor (ω) and dipole moment constitute the important basic properties. These are used to calculate various thermo-physical properties involved in governing equations. A vast database reporting these properties as well as, various methodologies used to evaluate several thermo-physical properties are reported in Reid et al. (R. C. Reid, J. M. Prausnitz and B. E. Poling, The properties of gases and liquids, third edition, McGraw Hill, 1977). The parameters affecting the calculation of various properties are outlined in this section.

Viscosity of a pure component (or individual species) is calculated as a function of its molecular weight, critical volume, temperature, viscosity collision integral and additional correction factors, as required. For example, based on the method of Chung et al. (refer Reid et al.), the viscosity is calculated as,

$$\mu_i = 40.785 \frac{F_c (T.MW_i)^{0.5}}{V_c^{2/3} \Omega_v}$$

Here, μ is in micropoise, F_c is correction factor, calculated using acentric factor, dipole moment and critical properties, and Ω_v is the viscosity collision integral, calculated using reduced temperature $(T_r = T/T_c)$, expressed as,

$$\Omega_v = A(T^*)^{-B} + C \exp(-DT^*) + E \exp(-FT^*)$$

Here,

$$T^* = 1.2593 T_r$$

The constants A through F are available in Reid et al. The correction factor F_c is determined using,

$$F_c = 1 - 0.2756\omega + 0.059035 \left(131.1 \frac{\mu}{(V_c T_c)^{1/2}} \right)^4 + \kappa$$

The correction factor, κ, is applied for highly polar substances such as methanol and water. The dipole moment of nonpolar species such as N_2, O_2 and CO_2 is zero, and it is nonzero for polar species such as H_2O. Mixture viscosity is calculated using the method of Wilke, as a function of mole fractions of species in the mixture. It is written as,

$$\mu = \sum \frac{X_i \mu_i}{\Phi}$$

Here, Φ is an interaction parameter, expressed in terms of molecular weights and viscosities of individual species, written as,

$$\Phi = \sum X_i \frac{\left[1 + \left(\frac{\mu_i}{\mu_j}\right)^{0.5} \left(\frac{MW_j}{MW_i}\right)^{0.25}\right]^2}{[8(1 + MW_i/MW_j)]^{0.5}}$$

Thermal conductivity of each species (in W/m K) is often calculated as a function of its viscosity, molecular weight, universal gas constant and a factor that depends upon acentric factor, reduced temperature and specific heat at constant volume of the species. Specific heat at constant volume is calculated from specific gas constant and specific heat at constant pressure. Therefore, thermal conductivity of a component requires the values of viscosity and specific heat at constant pressure of that component. It is expressed following the method of Chung et al., as,

$$\lambda_i = 3.75 \frac{\psi \mu R_u}{MW_i}$$

Here,

$$\psi = 1 + \alpha \left[\frac{0.215 + 0.28288\alpha - 1.061\beta + 0.26665Z}{0.6366 + \beta Z + 1.061\alpha\beta}\right]$$

Here,

$$\alpha = \frac{c_v}{R} - \frac{3}{2}; \beta = 0.7862 - 0.7109\omega + 1.3168\omega^2; Z = 2 + 10.5T_r^2$$

Mixture thermal conductivity is calculated in the same way as that of mixture viscosity.

Specific heat at constant pressure (in kJ/kg K) and specific enthalpy (in kJ/kg) of a pure component are often calculated as a piecewise polynomial involving temperature.

$$c_{pi} = R_u\left(a_1 + a_2T + a_3T^2 + a_4T^3 + a_5T^4\right)/MW_i$$

$$h_i = R_uT\left(a_1 + \frac{a_2T}{2} + a_3\frac{T^2}{3} + a_4\frac{T^3}{4} + a_5\frac{T^4}{5} + \frac{b_1}{T}\right)/MW_i$$

Here, the coefficients are specified for a given range of temperature in two sets. McBride et al. (B. J. McBride, G. Sanford and M. A. Reno, Coefficients for calculating thermodynamic and transport properties of individual species, NASA TM-4153, 1993) present coefficients for calculating specific heat and enthalpies of several species. Mixture-specific heat and enthalpy are calculated using simple chain rule involving mass fractions of the components.

$$c_p = \sum Y_i c_{pi}; \, h = \sum Y_i h_i$$

The ordinary binary diffusion coefficient or the binary diffusivity between a pair of species i and j, D_{ij}, in cm^2/s, is calculated as,

$$D_{ij} = \frac{0.00266 T^{3/2}}{p W_{ij}^{0.5} \sigma_{ij}^2 \Omega_D}$$

Here, W_{ij} is a function of molecular masses of the pair of species, written as,

$$W_{ij} = 2 \left(\frac{1}{MW_i} + \frac{1}{MW_j} \right)^{-1}$$

In terms of characteristic length of the molecules of species i and j, (in Å), σ_{ij} is written as,

$$\sigma_{ij} = \frac{\sigma_i + \sigma_j}{2}$$

The dimensionless collision integral, Ω_D, is written in terms of characteristic Lennard–Jones energy (ε/k), as,

$$\Omega_D = \frac{A}{(T^*)^B} + \frac{C}{\exp(DT^*)} + \frac{E}{\exp(FT^*)} + \frac{G}{\exp(HT^*)}$$

Here,

$$T^* = \frac{kT}{\epsilon_{ij}} \text{ and } \frac{\epsilon_{ij}}{k} = \left(\frac{\epsilon_i}{k} \frac{\epsilon_j}{k} \right)^{0.5}$$

The values of ϵ/κ and σ are taken from Reid et al. and the constants are: $A = 1.06036$, $B = 0.1561$, $C = 0.193$, $D = 0.47635$, $E = 1.03587$, $F = 1.52996$, $G = 1.76474$ and $H = 3.89411$.

Calculation of thermal diffusion coefficient is quite involved. For ideal gas mixtures, Ramshaw (J. D. Ramshaw, Hydrodynamic theory of multi-component diffusion and thermal diffusion in multi-temperature gas mixtures, Journal of non-equilibrium thermodynamics, 18, 121–134, 1993) suggested an approximation, further simplified by Pope and Gogos (D. N. Pope and G. Gogos, A New Multi-component Diffusion Formulation for the Finite-Volume Method: Application to Convective Droplet Combustion, Numerical Heat Transfer, Part B Fundamentals, Part B, 48, 3:213–234, 2005). This is written in terms of mass fractions of species i and j, and binary diffusion coefficient for the pair i-j, D_{ij}, as,

$$\frac{MW_{mix}}{\rho MW_i} \sum_{j \neq i} \frac{1}{MW_j D_{ij}} (Y_j D_{T,i} - Y_i D_{T,j}) = \sum_{j \neq i} (\alpha_{ji} - \alpha_{ij})$$

Here,

$$a_{ij} = F_{ij} \left(MW_j^2 \sum_{k=1}^{N} \frac{Y_k}{MW_k} \sigma_{jk}^2 \sqrt{\frac{MW_j + MW_k}{MW_j MW_k}} \right)^{-1}$$

Here,

$$F_{ij} = \frac{1}{3} Y_i Y_j \sigma_{ij}^2 \sqrt{\frac{MW_i + MW_j}{MW_i MW_j}} \frac{MW_i MW_j}{(MW_i + MW_j)^2} \left(10 \Omega_{i,j}^{(1,1)^*} - 12 \Omega_{i,j}^{(1,2)^*} \right)$$

Here, $\Omega_{i,j}^{(1,1)^*}$ and $\Omega_{i,j}^{(1,2)^*}$ are the collision integrals, evaluated at $T_{ij} = T(k/\epsilon_{ij})$, as reported in Neufeld et al. (P. D. Neufeld, A. R. Janzen, and R. A. Aziz, Empirical Equations to Calculate 16 of the Transport Collision Integrals for the Lennard–Jones (12–6) Potential, J. Chem. Phys., vol. 57, 1100–1102, 1972). Here, ϵ_i/k and σ_i are calculated using critical volume (cm^3/mol) and critical temperature (K), respectively, written as,

$$\frac{\epsilon_{ij}}{k} = \sqrt{\frac{\epsilon_i \epsilon_j}{k^2}}; \frac{\epsilon_i}{k} = \frac{T_{c,i}}{1.2593}; \sigma_i = 0.809 V_{c,i}^{1/3}$$

The $N - 1$ equations involving $D_{T,i}$ are solved with a constraint that $\sum D_{T,i} = 0$. The Soret diffusion velocity is then evaluated using Eq. (8.6). Even though the above procedure is a simplified one for ideal gas mixtures, this involves solving a set of equations iteratively.

As an alternative, diffusion velocity due to thermal diffusion is calculated using thermal diffusion ratio (K_T), defined following Chapman and Cowling (S. Chapman and T. G. Cowling, The Mathematical Theory of Non-Uniform Gases Cambridge University Press, Cambridge, 1970). For k^{th} species, it is evaluated as,

$$K_{T,k} = \sum k_{T,kj}; k \neq j$$

Here,

$$k_{T,kj} = \frac{15\left(2A_{kj}^* + 5\right)\left(6C_{kj}^* - 5\right)\left(MW_k - MW_j\right)}{2A_{kj}^*\left(16A_{kj}^* - 12B_{kj}^* + 55\right)\left(MW_k + MW_j\right)} X_j X_k$$

Here,

$$A_{kj}^* = \frac{\Omega_{kj}^{(2,2)}}{2\Omega_{kj}^{(1,1)}}; \quad B_{kj}^* = \frac{5\Omega_{kj}^{(1,2)} - \Omega_{kj}^{(1,3)}}{3\Omega_{kj}^{(1,1)}}; \quad C_{kj}^* = \frac{\Omega_{kj}^{(1,2)}}{3\Omega_{kj}^{(1,1)}}$$

Here, Ω_{kj} are the collision integrals, taken from Monchick and Mason (L. Monchick and E. A. Mason, Journal of Chemical Physics, 35, p. 1676, 1961). After evaluating the value of thermal diffusion ratio of kth species, its diffusion velocity is calculated using the expression,

$$\vec{v}_{k,\text{diff},T} = -\frac{D_{km}}{X_k} K_{T,k} \frac{\nabla T}{T}$$

Finally, the thermal diffusion coefficient can be calculated in a simplified algebraic form, as reported in Ansys FLUENT manual. It is written as,

$$D_{T,i} = -2.59 \times 10^{-7} T^{0.659} \left[\frac{MW_i^{0.511} X_i}{\sum_{i=1}^{N} MW_i^{0.511} X_i} - Y_i \right] \left[\frac{\sum_{i=1}^{N} MW_i^{0.511} X_i}{\sum_{i=1}^{N} MW_i^{0.489} X_i} \right]$$

8.2 Boundary Conditions

Boundary conditions for a reactive flow involve appropriate mathematical relations at the inlets, outlets, free boundaries, walls and interfaces, specified for all the variables such as velocity, pressure, temperature and species mass fractions.

Inlet At the inlets, specified velocity or mass flow rate or pressure can be specified along with the temperature. Velocity and mass flow rate are the quantities associated with the mixture. The mass fractions of $N - 1$ species in a mixture having N species are specified at this boundary. The mass fraction of the Nth species, which is usually the inert species present in large quantity, such as N_2, is calculated using the identity, $\sum Y_i = 1$. Based on the specified temperature and species concentrations, thermophysical properties, diffusion velocities of all species are calculated. If mass flow rate is given as input, using the mixture density and area of cross section, the mixture velocity is calculated. The mass flow rate or velocity, being vector quantities, proper direction cosines are also given as input, in order to resolve them in the required directions. In several cases, the direction of the incoming flow is the normal direction to the boundary. If pressure is specified in the inlet, based on the pressure gradient at the inlet, the velocity magnitude is determined using Bernoulli's equation locally, assuming that the flow coming in is a freestream (almost inviscid). The direction of this flow is perpendicular to the boundary.

Outlet In the case of confined/internal flows, gross flow leaves the domain out of this boundary. The pressure of the zone attached to this boundary, which may be even the atmosphere, is specified. All the variables are extrapolated from the interior, based on their gradients approaching a value of zero.

Free boundary This boundary is also called a *farfield* and represents a boundary usually open to the atmosphere. Atmospheric pressure is specified at this boundary. This boundary is chosen at a location, where the gradient of any variable normal to this boundary is quite small. Based on the pressure gradient, the flow can go out of the domain or it may come in. All the flow variables are extrapolated from the interior cells for the flow going out of the domain. For an incoming flow at the free boundary, velocity is calculated using the pressure gradient at this boundary. Since atmospheric air comes into the domain from atmosphere, the mass fractions of oxygen and nitrogen are specified as 0.23 and 0.77, respectively. The value of ambient temperature (300 K) is also specified.

Wall No-slip condition is specified for velocities at the wall. The fluid in contact with the wall will move with the velocity of the wall (this is usually zero). The first derivative of mass fractions of all the species normal to this boundary is specified as zero. Condition for temperature is specified either as fixed temperature (Dirichlet) or in the flux form (Neumann). If the wall is adiabatic, the first derivative of the temperature normal to this boundary is specified as zero. In many cases, based on the physical conditions, a solid domain is included adjacent to a part of the fluid domain, and heat transfer in the solid is solved. The solid domain may be the wall of the burner and the heat interaction to the wall is extremely important in predicting the flame characteristics, including its stability. Here, conjugate heat transfer occurs between solid and fluid domains.

Interface This is the boundary separating a condensed fuel surface and a gas phase. Condensed fuel may be a solid or liquid fuel. The gas-phase transport occurs at a much faster rate when compared to the "gasification" of the fuel from the condensed phase. The rate-limiting step is therefore, the gasification process, which is *vaporization* in the case of a liquid fuel, and *pyrolysis* in the case of a solid fuel.

A simplistic case is "steady" vaporization of a liquid fuel. Here, the fuel level is maintained constant by feeding the liquid fuel at the same rate as it vaporizes. An example for this scenario is the burning of thin fuel layer over the inert porous material such as a wick. Here the liquid-phase transport is not solved; however, the coupled interface boundary conditions are used to model the transport of heat and mass between two phases. A non-regressing liquid fuel surface with coupled interface conditions is specified based on the conservations of mass, momentum and energy. The equations are given as follows:

(i)   The liquid surface temperature, T_s, is calculated by considering conduction heat transfer from the ambient to the fuel surface (denoted by subscript s). The heat addition due to the radiation has not been considered here, but can be added along with suitable inputs for emissivity. In general, the heat flux from the gas phase to liquid phase is used for two purposes—sensible heating of the liquid to increase its temperature and supplying the latent heat to enable evaporation of the liquid from its surface. For the case of steady evaporation, a local thermodynamic equilibrium is achieved under the steady state, just like in the measurement of wet bulb temperature. Here, the heat from the gas phase is supplied as the latent heat. The energy balance at the interface, written in terms of mixture density, ρ, normal component of velocity, v (also called Stefan velocity), mixture thermal conductivity, λ, normal temperature gradient at the surface and latent heat of vaporization of the liquid fuel, h_{fg}, is given as,

$$\rho_s v_s h_{fg} = \lambda_s \left. \frac{\partial T}{\partial n} \right|_s \tag{8.17}$$

Here, subscript "s" denotes the quantity at the interface. Generally, the temperature gradient is expressed using surface temperature and temperature values in two adjacent interior nodes in the gas phase, using a 3-point interpolation.

(ii) The partial pressure, p, of the fuel vapor above the fuel surface is calculated as the saturation pressure at the surface temperature using the Antoine equation. Including fuel-specific constants, A, B and C, the Antoine equation is written as,

$$\log_{10} p = A - B/(C + T_s - 273.15)$$

There are other thermodynamic relationships, such as Clausius–Clapeyron equation, which can be used to calculate the partial pressure of the vapor over the interface, based on the surface temperature. The fuel mass fraction (Y_F) can be further evaluated from the partial pressure (or the mole fraction) of the species, its molecular mass and the mixture molecular mass.

(iii) The evaporation mass flux of the fuel, as calculated from Fick's law of ordinary diffusion, considering the diffusion coefficient between the fuel vapor and the mixture, D_{Fs}, and the normal gradient of fuel mass fraction at the surface are expressed as,

$$\rho_s v_s = \rho_s v_s Y_{Fs} - \rho D_{Fs} \left. \frac{\partial Y_F}{\partial n} \right|_s$$

The normal derivative of mass fraction of the fuel at the interface (denoted by subscript s) is calculated using 3-point interpolation or any other discretization schemes.

(iv)   The mass fractions of all species other than fuel, Y_i, are calculated by using Fick's law, with each of their net mass flux equated to zero. This is written as,

$$\rho_s v_s Y_i = \rho D_{is} \left. \frac{\partial Y_i}{\partial n} \right|_s$$

The normal derivative of mass fraction of species i at the interface (subscript s) is calculated in a manner similar to that of fuel. For each species, i, the diffusion coefficient with respect to its diffusion into the mixture, D_{is}, is calculated. The above equations are solved iteratively.

In the case of unsteady evaporation, as observed when a flame spreads over a liquid fuel surface, usually liquid-phase heat transfer has to be considered. The sensible heat supplied to the condensed phase becomes an important component. In several scenarios, the liquid-phase motion triggered by shear stress of the gas phase through the interface is considered to predict the evaporation rate properly. The shear stress term also includes the gradient of surface tension and the fluid flow triggered by the surface tension gradient, caused by the temperature gradient and/or the mass fraction gradient (in the case of multiple liquids in condensed phase), is called Marangoni convection.

In the case of solid fuels, the rate at which the fuel is pyrolyzed depends upon the temperature much strongly. Therefore, an Arrhenius-type equation is written to evaluate the mass flux of the pyrolysate (pyrolysis gases) from the solid surface, which is written as,

$$\dot{m}''_p = A \exp\left(-\frac{E_a}{R_u T_s}\right)$$

The surface temperature is evaluated using the gas-phase conductive heat flux at the interface, written as,

$$\dot{m}''_p \Delta h_P = \lambda_s \left. \frac{\partial T}{\partial n} \right|_s$$

Here, Δh_P is the heat of pyrolysis of the solid fuel. The mass fractions of the species are evaluated as done for the liquid fuel. As in the liquid fuel, in-depth heat conduction has to be solved inside the solid fuel in order to have proper energy balance.

8.3   A Note on Radiation and Soot Sub-Models

Radiation in flames and furnaces is an important phenomenon. Since a participating media, hot gases of combustion are involved, radiation calculations become complex. For a good review of flame radiation, an article by Sen and Puri (https://doi.org/10.

2495/978-1-84564-160-3/08) can be referred. This is due to non-gray nature of the gases. In particular, gases such as CO_2 and H_2O and solid particles such as soot, emit and absorb energy through radiation. In order to properly estimate the radiation involving these non-gray substances, radiation properties have to be known. This is the difficult part in radiation modeling. The spectral intensity of radiation at a given wavelength (λ) is obtained by solving Radiation Transfer Equation (RTE). This involves absorbing, emitting and scattering terms. Approaches in solving RTE include Monte Carlo methods, zonal methods, flux-based methods (discrete ordinates model), moment methods and spherical harmonics-based approximation (PN methods). After determining the spectral intensity, it is substituted in the gradient of the heat flux ($\nabla.q$) term that goes into the energy equation. This gradient is written in terms of G_λ, the incident radiation, spectral intensity of a blackbody at the given wavelength and absorption coefficient (α). Determination of radiation properties requires models. Detailed models involve integration of RTE over the entire molecular spectrum of gases. These are called line-by-line models and are computationally expensive. The next levels of models are called band models. The whole radiation spectrum is grouped into a number of bands. In each band, average radiation property, such as absorption coefficient, is calculated. These are often called narrow band models such as statistical narrow band, wide narrow band and so on. The last level of model is the volumetric model. Several studies have shown this model to predict flame temperature with reasonable accuracy and with minimum computational expense. Here, the radiation absorbed by participating gases such as CO, CO_2, CH_4 and H_2O is accounted for by calculating an average absorption coefficient, called Planck's mean absorption coefficient (α_P). This is calculated as a function of partial pressures (p_i) of the participating species and the temperature. For non-luminous flames, in which soot formation is negligible, the optical thickness is quite less. Therefore, based on optically thin approximation, the volumetric radiation term ($\dot{q}_R'''$) in W/m^3 is calculated as,

$$\dot{q}_R''' = 4\sigma \left(T^4 - T_\infty^4 \right) \sum_i (p_i \alpha_{Pi})$$

Here, the α_{Pi} is the absorption coefficient of species i calculated at the temperature T. Details of the calculation of Plank's mean absorption coefficient is available in Barlow et al. (R. S. Barlow. A. N. Karpetis, J. H. Frank and J. Y. Chen, 2001, Scalar profiles and NO formation in laminar opposed-flow partially premixed methane/air flames, Combustion and Flame, 127, 2102–2118). This volumetric term is given as a negative source to the energy equation. This model can be extended to include radiation due to soot particles, if soot volume fraction is known.

Soot formation is also a complex process to model. A soot particle is formed by a set of species, called *soot precursors*. Polycyclic Aromatic Hydrocarbons (PAH) having cyclic bond structure are formed during pyrolysis (disintegration) of the fuel. Compounds such as acetylene (C_2H_2) and ethylene (C_2H_4) participate in the reactions to form some amount of PAH such as C_6H_6. The formation of soot precursors depends upon the fuel structure and the flame temperature. The PAH molecules *nucleate* soot particles and number of such particles increases, resulting in higher *number density*. These particles undergo *surface growth* due to the removal of hydrogen along with an addition of acetylene. The increase in the number density facilitates increase of the size of the particles, which is termed as *agglomeration*, where smaller particles coagulate to form large-chain structures. The soot particles undergo *oxidation* where the carbon is oxidized by oxygen, dependent on temperature and oxygen availability.

In several cases, soot modeling has to be included. Soot models can be classified as empirical, semiempirical and theoretical models. Empirical soot models involve experimental data-based curve fit values for soot formation and oxidation rates. Here, the rate of soot formation is written in an Arrhenius form involving mass of the fuel, pre-exponential factor and an activation energy. Soot oxidation rate is similarly written in terms of mass of the soot, mole fraction of oxygen and a different set of pre-exponential factor and an activation energy values. A mechanism consisting of the following global reactions is included in semiempirical models:

$$C_2H_2 \rightarrow 2C(s) + H_2 \,(\text{nucleation from a precursor})$$

$$C_2H_2 + n\,C(s) \rightarrow (n+2)C(s) + H_2 \,(\text{surface growth})$$

$$C(s) + 0.5\,O_2 \rightarrow CO \,(\text{soot oxidation})$$

$$n\,C(s) \rightarrow C_n(s) \,(\text{agglomeration})$$

Rates of these reactions are evaluated for a particular combustion system. Semiempirical soot models, which solve for differential equations involving soot number density and soot volume fraction, are also reported in the literature. Here, the nucleation, coagulation or agglomeration, surface growth and oxidation steps are included as appropriate source terms. More detailed models, such as nine-steps model, are also available to model soot formation. Soot surface oxidation involves its reaction with OH and O_2. Soot models available in commercial CFD codes account for all the four processes based on the empirical correlations for a given configuration, fuel and mode of combustion. Thus, these are mostly suitable for practical engineering calculations and not for absolute prediction of soot volume fraction in a flame. The soot model parameters have to be calibrated for a given fuel and oxidizer combination and must be validated with appropriate experimental data. With the prediction of soot volume fraction (f_v), the radiation due to soot particles, in its simplistic form as a volumetric term, is written as,

$$\dot{q}_R''' = 4\sigma\left(T^4 - T_\infty^4\right) \sum_i (p_i \alpha_{Pi}) + Cf_v\left(T^5 - T_\infty^5\right)$$

Here, C is a constant related to power density in W/m^3 [F.S. Liu, J.L Consalvi, A. Fuentes, "Effects of water vapor addition to the air stream on soot formation and flame properties in a laminar co-flow ethylene/air diffusion flame," Combustion and Flame 161 (2014): 1724–1734].

8.4 Selection for Mesh and Time Step

8.4.1 Grid Independence Study

A numerical solution is obtained at finite number of points in the domain. However, the choice of the number of points should not alter the nature and accuracy of the solution. The mesh should have sufficient number of points that can predict all the features of the solution. For example, the maximum value and its location in a profile of a variable such as temperature, which includes a strong gradient, should be predicted well, and this should not change when the number of cells is further increased. This is achieved by conducting a grid independence study, and the methodology adopted for this is given below.

(1) The domain is first discretized using a *coarse grid*, having a certain number of cells. This may be a non-uniform grid, with finer cells at the walls and other places where higher gradients are expected.

(2) A case is chosen to conduct the grid independence study. This may be a validation case, for which experimental data are available for comparison. It may also be a case within the parametric matrix used to conduct the parametric study, which involves maximum Reynolds number, maximum heat release rate and so on.

(3) Simulation of the chosen case is carried out using the coarse grid. Sufficient number of iterations are executed till convergence. Convergence is checked by monitoring the change in the normalized value of the variable at the current iteration as compared to that in the previous iteration. Further, mass balance is checked using the ratio of net efflux of the mass to the least mass flux entering the domain. This quantity is ensured to be less than 1%. Similarly, energy and species atom balances are checked. The achievement of steady state is also confirmed by plotting the profiles of temperature, major and minor species at several iterations and observing negligible variation in the profiles with iterations. In the case of an unsteady solution, convergence is ensured in each time step and time marching is done for the required time period. The time period is usually chosen based on the *flow time* associated with the problem. Transient data are time-averaged for a given time period to obtain the time-averaged quantities.

(4) After satisfactory convergence of the simulation with coarse grid, the gradients of variables such as temperature, reactant and product species, and radicals are checked. Based on the gradients in these quantities, the number of cells at selected locations are increased in order to resolve the gradients well. An *intermediate grid* is formed in this manner. The same case is simulated using

the intermediate grid until satisfactory convergence for the steady solution or required time marching is achieved for the transient case.

(5) Similar to step 3, the gradients are checked and a further increase in the number of cells is made to arrive at the *fine grid*. The same case is executed with the fine grid.

(6) The results from the three grids, coarse grid, intermediate grid and fine grid, are compared by analyzing the profiles of temperature, species mass fraction and velocity, for example. If the differences between the intermediate and fine grids are low and within a given percentage (say 5%), then the intermediate grid can be considered for further simulations. If the differences are more than the given percentage, then a *finer grid* with more cells than the fine grid is formed and the above procedure is repeated. The aim is to arrive at a mesh with sufficient number of cells that will provide results, which will not significantly vary further.

8.4.2 Time-Step Selection

A reactive flow involves timescales as low as the reaction times. In transient processes such as ignition, extinction, flame spread and so on, the timescales are of the order of milliseconds. If a reaction mechanism is involved in the simulations, the fastest reaction will have the lowest time. The time step should be able to capture such fast reactions. A time step for reactive flow calculations is usually fixed by estimating the characteristic time for a chain propagating or chain branching reactions, which are the faster reactions. Several researchers have used time steps of the order of 10^{-4} s or 10^{-5} s. Similar to a grid independence study, a time-step independence study is also conducted. Here, the simulation is carried out by using a time step of a given value and the same case is simulated with a time step one order of magnitude lower than the earlier value. If the differences in the results obtained from these two time steps are not significant, then the higher time step is used.

8.5 Case Study

In this section, simulation of a laminar diffusion (non-premixed) flame of LPG in an open-air environment at atmospheric and under normal gravity condition is presented as a case study. An axisymmetric domain is considered to simulate the flame from a cylindrical burner (Fig. 8.1). Domain extents are kept as 24d in the axial direction from the burner exit and 5d (radius) in the radial direction from the axis. Here, d is the internal diameter of the burner (10 mm). The boundary conditions are also shown in Fig. 8.1. At the axis, the normal velocity component is set to zero and first derivatives of all other variables in the radial direction are set to zero. Free boundaries are specified conditions as discussed in Sect. 8.2. The wall of the burner

Fig. 8.1 Schematic of the
computational domain with
boundary conditions

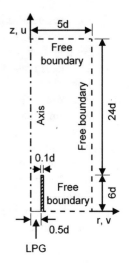

has been meshed and conjugate heat transfer is solved. LPG is a multi-component fuel and has a typical composition of 0.03% CH_4, 0.96% C_2H_6, 13.31% C_3H_8, 10.22% C_3H_6, 30.23% $i\text{-}C_4H_{10}$, 25.32% $n\text{-}C_4H_{10}$, 3.98% C_4H_8, 5.03% $i\text{-}C_4H_8$, 4.99% $trans\text{-}2\text{-}C_4H_8$, 3.64% $cis\text{-}2\text{-}C_4H_8$, 1.96% $i\text{-}C_5H_{12}$ and 0.33% $n\text{-}C_5H_{12}$. Therefore, to model its oxidation, a short chemical kinetic mechanism, consisting of 43 species and 392 reactions, has been used. Full multi-component diffusion including thermal diffusion has been considered. Temperature- and concentration-dependent thermo-physical properties, SIMPLE algorithm for pressure–velocity coupling and second-order upwind schemes have been employed. To ensure convergence, normalized residual value is ascertained to be less than 1×10^{-3} for continuity and momentum, and less than 1×10^{-6} for energy, within each time step. A simplified soot model has been used to predict soot volume fraction considering C_2H_2 and C_2H_4 as soot precursors. Combined gas and soot radiation model, as discussed in Sect. 8.3, has been used as a volumetric sink term in energy equation.

First, results from grid independence and time-step independence studies are shown. Fuel flow rate is kept as 10 g/h. Due to the momentum of the fuel jet, it entrains atmospheric air and mixes. Upon ignition using a small zone of high temperature of 1500 K near the burner exit, a flame is formed anchored close to the burner exit. This case has been simulated using three grids with number of non-uniformly placed cells varying as 13,440 (coarse grid), 51,360 (intermediate grid) and 115,590 (fine grid). The case is transient in nature. Simulations have been carried out using two values of time step—10^{-4} and 10^{-5} s.

Figure 8.2 shows the results from three grids in terms of axial and radial direction profiles of temperature, velocity and mass fractions of species. Temperature and velocity are normalized with their corresponding maximum values along the axis for axial profiles and with their maximum values along the radial direction at a

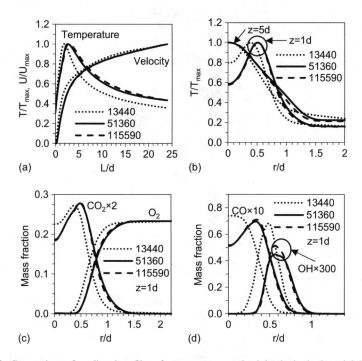

Fig. 8.2 Comparison of predicted profiles of **a** temperature and axial velocity in the axial direction, **b** temperature and mass fractions of **c** O_2, CO_2 and **d** CO, OH in the radial direction in a LPG–air jet diffusion flame for a fuel flow rate 10 g/h using three grids

given axial position in the radial profiles. The axial and radial distances have been normalized with respect to burner diameter. It is clear that a significant difference prevails between the coarse and intermediate grids, and the differences between the intermediate and fine grids are negligible in all profiles. Therefore, the intermediate grid with 51,360 cells is good enough for further calculations. Similarly, Fig. 8.3 displays results from time-step independence study, showing that negligible differences prevail between the results got from two time steps, 10^{-4} and 10^{-5} s. Thus, a time step of 10^{-4} s can be used further.

Fig. 8.3 Comparison of predicted species profiles in the radial direction in a transient simulation using time-step values of 10^{-4} and 10^{-5} s

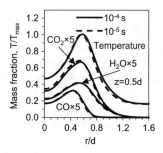

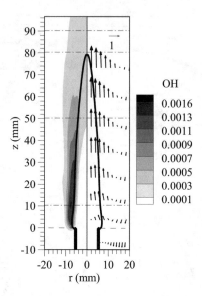

Fig. 8.4 Contours of OH mass fraction (greyscale) with stoichiometric contour line (black line) and velocity vectors in LPG–air jet diffusion flame for a fuel flow rate 22.5 g/h

Figure 8.4 presents the greyscale contours of OH mass fraction, stoichiometric contour line, indicating the location where the fuel species and oxygen mix at stoichiometric proportions, and the vectors of the mixture velocity. The stoichiometric contour line indicates the flame (primary reaction) zone. It is clear that maximum OH is present along this line. The velocity is seen to increase across this time indicating that hot gases are accelerated in the upward direction due to buoyancy, creating a recirculation zone adjacent to the flame surface showing the air entrainment. Figure 8.5 presents the structure of the flame in terms of axial and radial profiles of temperature and species mass fractions.

The axial profile of fuel (Fig. 8.5a) shows that the fuel (sum of mass fractions of all fuel components in the fuel mixture) decreases to almost zero value at the axial location of 8d from the burner exit. This represents the flame tip. The velocity shows a monotonically increasing trend along the axis due to acceleration due to buoyancy-induced force. Temperature increases along the axis and reaches its maximum value around the flame tip, and it decreases subsequently. Radial profiles of temperature (Fig. 8.5b) show a double peak profile displaying a peak away from the axis at axial locations below the flame tip. At the flame tip, temperature presents its maximum value at the axis. Radial profiles of major reactant and product species are shown in Fig. 8.5c. Radial locations where O_2 decreases to an almost zero value or where CO_2 peaks represent the radius of the flame at that axial location. It is found to coincide with stoichiometric contour line. At the flame tip, CO_2 produces its maximum at the axis and O_2 penetrates until the axis. Radial profiles of minor species are presented in Fig. 8.5d. Peak value of CO decreases with increase in the axial location where

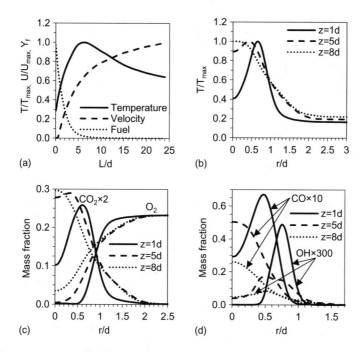

Fig. 8.5 Comparison of predicted profiles of **a** temperature, velocity and mass fraction of fuel in the axial direction, **b** temperature and mass fractions of **c** O_2, CO_2 and **d** CO, OH in the radial direction at $z = 1d$, 5d and 8d from burner exit, in LPG–air jet diffusion flame for a fuel flow rate 22.5 g/h

the radial profile has been plotted. At the flame tip some amount of CO is present, which may also be contributed by the dissociation of CO_2. Contours of mass fraction of OH also show similar trends.

Figure 8.6 shows axial and radial profiles of mass fraction of C_2H_2 (one of the soot precursor) and soot volume fraction. It is clear that maximum value of soot volume fraction occurs just above the location of the maximum mass fraction of C_2H_2 along the axis (Fig. 8.6a, b).

Soot is subsequently oxidized such that the soot volume fraction decreases along the axis after reaching a peak. Radial profiles of C_2H_2 mass fraction (Fig. 8.6c) indicate that its maximum occurs at the axis at an axial location of around 5d and in the locations below that it produces a peak around the maximum temperature zone. Radial profiles of soot mass fraction clearly show that soot is not formed at primary oxidation zones ($z = 1d$), and it forms only after enough precursors are formed. An increase in the soot oxidation is indicated by the decrease in the soot volume fraction at higher axial locations both in axial and radial profiles.

Review Questions

1.   Write an expression to calculate the species velocity.

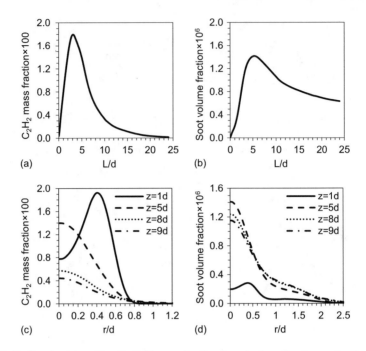

Fig. 8.6 Predicted axial profiles of **a** C_2H_2 mass fraction, **b** soot volume fraction, and radial profiles at $z = $ 1d, 5d, 8d and 9d, of **c** C_2H_2 mass fraction, **d** soot volume fraction in LPG–air jet diffusion flame for a fuel flow rate 22.5 g/h

2. What is the constraint in relating the diffusion fluxes?
3. Write the expression used to estimate the diffusion coefficient of a species into a mixture.
4. Why a correction is applied to diffusion velocity? What is the constraint used in that?
5. What is the unit for thermal diffusion coefficient?
6. Write the species conservation in spherical coordinates.
7. Relate stress tensor and rate of strain in Cartesian coordinates.
8. What is Dufour effect?
9. Write the energy equation in terms of enthalpy in cylindrical polar coordinates.
10. Write the energy diffusion due to species diffusion in terms of temperature.
11. How are boundary conditions prescribed for species at the walls?
12. Write the energy balance at the liquid–vapor interface for steady vaporization.
13. How is pyrolysis of solid fuel modeled?
14. What are the factors affecting the absorption coefficient of a gas mixture?
15. Write the important reactions constituting soot formation.
16. How is mass imbalance calculated?
17. How is atom balance calculated in the domain?
18. What is the typical time step used in reacting flows?

References

1. **Bernard Lewis** and **Guenther Von Elbe**, *Combustion, Flames, and Explosions of Gases*, 1951, Academic Press.
2. **Marion L. Smith** and **Karl W. Stinson**, *Fuels and Combustion*, 1952, McGraw-Hill, Inc., USA.
3. **D. B. Spalding**, *Some Fundamentals of Combustion*, 1955, Academic Press Inc., New York.
4. **Forman A. Williams**, *Combustion Theory*, 1965, Addison-Wesley Pub. Co., 2nd edition, 1985, ABP, Perceus Books Publishing, USA.
5. **Roger A. Strehlow**, *Fundamentals of Combustion*, 1968, International Textbook Co. *Combustion Fundamentals*, International Edition, 1985, McGraw-Hill, Inc., USA.
6. **Murthy A. Kanury**, *Introduction to Combustion Phenomena*, 1975, Gordon and Breach Science Publishers, Inc., New York.
7. **Irvin Glassman**, *Combustion*, 1977, 1987, 1996, Academic Press, USA. **Irvin Glassman** and **Richard Yetter**, *Combustion*, 2008, Academic Press, USA.
8. **D. B. Spalding**, *Combustion and Mass Transfer*, 1979, Pergamon Press, UK.
9. **Kenneth K. Kuo**, *Principles of Combustion*, 1986, 2nd edition, 2005, John Wiley, USA.
10. **Alan Williams**, *Combustion of Liquid Fuel Sprays*, 1990, Butterworths (Canada) Limited.
11. **J. Warnatz, U. Mass** and **R. W. Dibble**, *Combustion*, 1996, Springer-Verlag, Germany.
12. **Stephen R. Turns**, *An Introduction to Combustion*, 2nd edition, 2000, 3rd edition, 2012, McGraw-Hill, Inc., USA.
13. **Norbert Peters**, *Turbulent Combustion*, 2000, Cambridge University Press, UK.
14. **Thierry Poinsot** and **Denis Veynante**, *Theoretical and Numerical Combustion*, 2005, R.T. Edwards, Inc., USA.
15. **Kalyan Annamalai** and **Ishwar K. Puri**, *Combustion Science and Engineering*, 2006, CRC Press, USA.
16. **Prabir Basu**, *Combustion and Gasification in Fluidized Beds*, 2006, CRC Press, USA.
17. **H. S. Mukunda**, *Understanding Combustion*, 2nd edition, 2009, University Press, India.
18. **Chung K. Law**, *Combustion Physics*, 2010, Cambridge University Press, UK.
19. **Arthur H. Lefebvre** and **Dilip R. Ballal**, *Gas Turbine Combustion*, 3rd edition, 2010, CRC Press, USA.
20. **Kenneth W. Ragland** and **Kenneth M. Bryden**, *Combustion Engineering*, 2011, CRC Press, USA.

Suggested Reading

1. **Bernard Lewis** and **Guenther Von Elbe**, *Combustion, Flames, and Explosions of Gases*, 1951, Academic Press.
2. **Marion L. Smith** and **Karl W. Stinson**, *Fuels and Combustion*, 1952, McGraw-Hill, Inc., USA.
3. **D. B. Spalding**, *Some Fundamentals of Combustion*, 1955, Academic Press Inc., New York.
4. **Forman A. Williams**, *Combustion Theory*, 1965, Addison-Wesley Pub. Co., 2nd edition, 1985, ABP, Perceus Books Publishing, USA.
5. **Roger A. Strehlow**, *Fundamentals of Combustion*, 1968, International Textbook Co. *Combustion Fundamentals*, International Edition, 1985, McGraw-Hill, Inc., USA.
6. **Murthy A. Kanury**, *Introduction to Combustion Phenomena*, 1975, Gordon and Breach Science Publishers, Inc., New York.
7. **Irvin Glassman**, *Combustion*, 1977, 1987, 1996, Academic Press, USA. **Irvin Glassman** and **Richard Yetter**, *Combustion*, 2008, Academic Press, USA.
8. **D. B. Spalding**, *Combustion and Mass Transfer*, 1979, Pergamon Press, UK.
9. **Kenneth K. Kuo**, *Principles of Combustion*, 1986, 2nd edition, 2005, John Wiley, USA.
10. **Alan Williams**, *Combustion of Liquid Fuel Sprays*, 1990, Butterworths (Canada) Limited.
11. **J. Warnatz, U. Mass** and **R. W. Dibble**, *Combustion*, 1996, Springer - Verlag, Germany.
12. **Stephen R. Turns**, *An Introduction to Combustion*, 2nd edition, 2000, 3rd edition, 2012, McGraw-Hill, Inc., USA.
13. **Norbert Peters**, *Turbulent Combustion*, 2000, Cambridge University Press, UK.
14. **Thierry Poinsot** and **Denis Veynante**, *Theoretical and Numerical Combustion*, 2005, R.T. Edwards, Inc., USA.
15. **Kalyan Annamalai** and **Ishwar K. Puri**, *Combustion Science and Engineering*, 2006, CRC Press, USA.

16. **Prabir Basu**, *Combustion and Gasification in Fluidized Beds*, 2006, CRC Press, USA.
17. **H. S. Mukunda**, *Understanding Combustion*, 2nd edition, 2009, University Press, India.
18. **Chung K. Law**, *Combustion Physics*, 2010, Cambridge University Press, UK.
19. **Arthur H. Lefebvre** and **Dilip R. Ballal**, *Gas Turbine Combustion*, 3rd edition, 2010, CRC Press, USA.
20. **Kenneth W. Ragland** and **Kenneth M. Bryden**, *Combustion Engineering*, 2011, CRC Press, USA.

Index

Printed in the United States
by Baker & Taylor Publisher Services